ADVANCE PRAISE FOR

Writing Critically in STEAM

"In Writing Critically in STEAM, Ness is again concise yet expansive. In many ways, considering his career, he was destined to write this book. On one level this book is about how to write in the different STEAM disciplines. We know intuitively that we write differently in math, for example, than in technology, but we don't always know what the large differences are—not to mention the subtle gray zones between the genres. As Ness teases out these differences, he explores the multiple underlying dynamics of writing that take place before, during and even after the writing process within STEAM. This focus includes the writer's stance in relation to thinking, working, and communicating within the different disciplines. It includes subtle differences between the genres within the disciplines, such as those between scientific writing and science writing. And it includes the formation of representation within the disciplines. "
—Richard D. Sawyer Professor, Washington State University, Vancouver

"Daniel Ness reminds us that in order to not have writing criticized, we attend to the verity within STEM. As in all disciplines, writing critically is essential. Not affirming excuses made by scientific writers that due to the discipline, excellence in writing is not expected, the book advises us that we develop appropriately when we have learned to write. The significance is clear, STEM scholars and students should understand and exercise excellence and criticality in their writing, bottom line. Writing Critically is STEM's just do it! clarion call. "
—Shirley R Steinberg, PhD Editor; Sage Handbook of Critical Pedagogies Shirley R. Steinberg Professor, University of Calgary

Writing Critically in STEAM

Critical Literacies and Language

Brett Elizabeth Blake and Judith M. Dunkerly
Series Editors

Vol. 1

Daniel Ness

Writing Critically in STEAM

PETER LANG
New York · Berlin · Bruxelles · Chennai · Lausanne · Oxford

Library of Congress Cataloging-in-Publication Data

Names: Ness, Daniel, author.
Title: Writing critically in STEAM / Daniel Ness.
Description: New York : Peter Lang, [2024] | Series: Critical literacies and language, 2993-9488 ; vol. 1 | Includes bibliographical references and index.
Identifiers: LCCN 2023033536 (print) | LCCN 2023033537 (ebook) | ISBN 9781636673226 (paperback) | ISBN 9781636673202 (pdf) | ISBN 9781636673219 (epub)
Subjects: LCSH: Technical writing. | Mathematics–Authorship | Architectural writing. | Critical theory.
Classification: LCC T11 .N47 2024 (print) | LCC T11 (ebook) | DDC 808.06/66–dc23/eng/20231026
LC record available at https://lccn.loc.gov/2023033536
LC ebook record available at https://lccn.loc.gov/2023033537

DOI 10.3726/b21092

Bibliographic information published by the Deutsche Nationalbibliothek.
The German National Library lists this publication in the German National Bibliography; detailed bibliographic data is available on the Internet at http://dnb.d-nb.de.

Cover design by Peter Lang Group AG

ISSN 2993-9488 (print)
ISSN 2993-9496 (online)
ISBN 9781636673226 (paperback)
ISBN 9781636673202 (ebook)
ISBN 9781636673219 (epub)
DOI 10.3726/b21092

© 2024 Peter Lang Group AG, Lausanne
Published by Peter Lang Publishing Inc., New York, USA
info@peterlang.com - www.peterlang.com

All rights reserved.
All parts of this publication are protected by copyright.
Any utilization outside the strict limits of the copyright law, without the permission of the publisher, is forbidden and liable to prosecution.
This applies in particular to reproductions, translations, microfilming, and storage and processing in electronic retrieval systems.

This publication has been peer reviewed.

To Steve Farenga – A great friend, colleague, and mentor

CONTENTS

Foreword		ix
Acknowledgments		xiii
Introduction		1
Chapter 1.	What Does it Mean to Write Critically in STEAM?	7
Chapter 2.	Tools for Writing Critically in STEAM	19
Chapter 3.	From Brainstorming to Writing Critically in STEAM	31
Chapter 4.	Writing Critically in the Natural Sciences	45
Chapter 5.	Writing Critically in Technology	63
Chapter 6.	Writing Critically in Engineering, Art, and Architecture	73
Chapter 7.	Writing Critically in Mathematics	91

| Chapter 8. | Questioning Norms in Critical STEAM Writing | 105 |

	Glossary of Technical Terms in Writing and STEAM	123
	Appendix A	129
	Appendix B	139
	Appendix C	159
	Appendix D	165
	References	173
	Index	179

FOREWORD

Foreword to Writing Critically in STEAM

Daniel Ness is that rare scholar who is a specialist generalist. By that I mean that he has worked deeply in a number of interdisciplinary fields. He not only has produced cutting-edge scholarship in mathematics education, psychology, art and architecture, music education, critical pedagogy and theory, school reform, and the foundations of education, but also has worked at the multiple and generative intersections of these fields. For over 20 years now, I've had the good fortune to work closely with and learn from him.

We met years ago as graduate students at Teachers College, Columbia University. For a couple of years, mentored by Dr. Gary Natriello, who was the editor of the *Teachers College Record*, he and I and a few others did the initial internal reviews for the submissions to the journal. At weekly meetings, we shared and critiqued our reviews. I recall then—and now as I've just finished reading his latest work—just how layered and complex his writing was and is. His manuscript submission reviews back then offered a critical examination of power dynamics, explored historical and contemporary debates within a range of fields, and produced and critiqued new knowledge within a wide range of disciplines. His reviews were informative. They were also pedagogical: they scaffolded our cognitive development and helped us to grow as thinkers and scholars. And, perhaps most importantly, they were accessible.

In *Writing Critically in STEAM*—his latest publication—he is again concise yet expansive. In many ways, considering his career, he was destined to write this book. On one level this book is about how to write in the different STEAM disciplines. We know intuitively that we write differently in mathematics, for example, than in technology, but we don't always know what the large differences are—not to mention the subtle gray zones between the genres. As Ness teases out these differences, he explores the multiple underlying dynamics of writing that take place before, during, and even after the writing process within STEAM. This focus includes the writer's stance in relation to thinking, working, and communicating within the different disciplines. It includes subtle differences between the genres within the disciplines, such as those between scientific writing and science writing. And it includes the formation of representation within the disciplines. Referencing Lev Vygotsky, he explores how the process of learning to write transcends the production of text: the concept of symbol formation within such writing leads development.

Examining how representation regulates and structures liberation, Ness highlights a central theme of this book: writing in STEAM can indeed be critical and challenge societal inequities. This critical stance in STEAM writing relates to form, the pronouns writers use, the examples they give, and the assumptions they make. But writing for liberation goes beyond these more formal considerations to include personal history, the value of voice, and the acknowledgement of the humanistic roots of science. Writing in STEAM can facilitate Paolo Freire's dialogic problem posing and critical pedagogy. As Ness states, "In institutionalizing the status quo, schools have paid virtually all its attention to the memorization of facts and little, if anything, to the development of critical literacy." Exploring the institutionalization of inequity, he asks a significant and provocative question: "Is adherence to national standards and writing critically in STEAM contradictory?"

In contrast to structures that alienate students from the writing process in STEAM, Ness invites them to find a voice and identity within the process. He states, "Be your own author. Indeed, emulating great models is a good thing, but it's important for readers to listen to your own voice on a given topic and not someone else's." Reading this book has motivated (and empowered) me to write outside my comfort zone in STEAM. For me, knowing how much I have learned from Dan Ness over the years, it was no surprise that I would continue to learn from this text (both about writing and about myself).

As a text that clarifies the layered process of writing in STEAM, this new book contributes greatly to scholarship. But his text goes beyond methods of writing to explore the critical dynamics of writing within STEAM and the intersections between structural equity and individual development.

Perhaps for me the highlight of this text was simply hearing Dan Ness's personal voice within his writing. You can clearly hear his voice in the following quote:

> You joined me on this voyage of STEAM writing at the beginning of the book, but the voyage does not end here. Rather, our investigation in STEAM writing should be the commencement of a life-long voyage, full of writing adventures that are convincing, persuasive, motivating, and exciting for your readers.

Reading this book was a rich journey for me. It took me through over 20 years of Dan Ness's scholarship (and friendship) to future possibilities of writing within STEAM for social justice.

Richard D. Sawyer, Professor
Department of Teaching and Learning
Washington State University, Vancouver

ACKNOWLEDGMENTS

There are several individuals who I cannot thank enough for helping me with this book project. Without their encouragement and support, much, if not all, of the words on these pages would be left void. First, I am grateful to the Peter Lang staff, Alison Jefferson and Joshua Charles in particular, for helping me see this project through from beginning to end. Readers should note that all errors that remain in this book are inadvertent and those entirely of my own doing. I, therefore, take full responsibility for any error or oversight that might be discovered herein. Next, a note of tribute goes to Brett Blake. It is due to her inspiration and assurance of my abilities that I carefully and thoughtfully spilt ink on these pages. Her expertise in critical literacy enabled me to connect critical pedagogy with STEAM and STEAM education. Special thanks go to Shirley Steinberg, Professor of Education at the University of Calgary, who read through these pages and provided extensive suggestions and comments. Rick Sawyer, Professor of Education at Washington State University, Vancouver, also read the manuscript from cover to cover and wrote a praiseworthy foreword to this book; one cannot ask for a more moving and inspiring endorsement! My intellectual conversations on everything about curriculum with Wanying Wang, Visiting Professor at St. John's University, serve as a beacon that profoundly has influenced me in many a direction. I cannot thank her enough. I am also indebted to my son, Eric, for the completion of this book; his continued interests in music, art, and STEAM subjects have inspired me tremendously. Lastly, this book is dedicated to my mentor, Stephen Farenga, teacher, researcher, Professor of Science Education at the City University of New York, Queens College, scholar, and friend, who, in large part, shaped me into who I am today.

INTRODUCTION

Writing can be a tortuous process in any discipline. However, it can seem particularly daunting in the fields of science, technology, engineering, the arts or architecture, and mathematics (henceforth, STEAM). In fact, most of the STEAM professionals with whom I have come into contact have said things like, "As an engineer, I write only when I have to . . ." or "engaging in science is easy for me but sitting down and writing my manuscripts is so painful" or "I'm really good at math but I'm a terrible writer . . ." But this does not have to be the case. Despite challenges along the way, critical writing for the STEAM learner, practitioner, and even researcher (yes, I said researcher!)—a group of individuals to whom I will henceforth refer as STEAMers—can become an easier process in STEAM especially when one's ideas are fleshed out and organized in a systematic manner.

The "A" in STEAM

While there is a rather large selection of chapters or articles on reading and writing in STEM, at present, to my knowledge, there are few, if any, readers or supplementary materials that serve to support researchers and related professionals to become better critical writers. To help fill this gap, *Writing Critically in STEAM* focuses primarily on what the scientist, technology specialist, engineer, artist, art historian, or architect, and mathematician need to know in order to become a better writer.

Because the arts play an intrinsically important and symbiotically related role in traditional STEM subjects, you will notice that the acronym STEAM is used throughout this book (instead of the more common acronym STEM). We need only think about the field of architecture and the architect's time-honored tradition of fulfilling three basic so-called virtues, also known as the Vitruvian trinity, of real-world constructions: *firmitas, utilitas,* and *venustas*.

In architecture, strength and durability is of fundamental importance. *Firmitas* is a structure's level of durability after extended use and exposure to the natural elements. Over the centuries, architects, engineers, natural scientists, and mathematicians have been able to calculate with greater levels of exactitude the expected life spans of their structures. While some of their findings were the result of testing and retesting, others, unfortunately, were not. In fact, most structural change considerations were due to accidents, natural disasters, and human-initiated disasters that necessitated researchers to find ways to address structural or mechanical failures in preventing such occurrences from reoccurring. More specifically, they found that certain materials have greater durability than others. Moreover, since ancient times, some materials such as marble, concrete, and brick, have had the illusion of strong durability. Eventually, these materials proved to be inferior to yet stronger materials like reinforced concrete and steel. From the perspective of the student or professional, *firmitas* clearly requires the expertise of mathematicians, physical and environmental scientists (Baker, 2017), mechanical engineers, and materials engineers.

The second virtue in the Vitruvian trinity, *utilitas* refers to a structure's ability to respond to the needs of its anticipated inhabitants or users. We see the word "utility" in the Latin *utilitas*. It emphasizes the importance of a structure's function. The architect's consideration of function can be applied to structures that are enclosed for human activity, such as schools, hospitals, homes, and skyscrapers, as well as those that are not, such as monuments, radio masts and towers, silos, and grain elevators. Here, the student or professional recognizes and appreciates the need for the expertise of mathematicians, building service engineers, electrical engineers, and mechanical engineers.

Venustas, the third virtue in the Vitruvian trinity, points not to strength or function, but to aesthetics and beauty of a structure. The origin of the word *venustas* itself emanates from the deity Venus, who has come down through the ages to signify beauty and aesthetic appeal. From time immemorial to contemporary society, *venustas* in architecture has placed strong emphasis on the association between the spatial and aesthetic conditions of a structure. These include, but are not limited to, proportion, scale, levels of light and shade, visual textures and structural patterns. The student or professional can appreciate the knowledge and foresight of experts in mathematics, natural science, and many branches of engineering, and indeed, architecture for integrating aesthetic value to structure. Simply consider the mathematical and scientific underpinnings of the Golden Ratio and its application to the construction of the Pyramids of Giza in Egypt, the Parthenon in Athens, Greece, the Great Mosque of Kairouan in Tunisia, the Temple of Kukulcan built in Mesoamerican Mexico, or the more contemporary CN Tower in Toronto. As observed in Figure I1, the key point I want to stress here is that the arts, in the broadest perspective, can enable writers to not only convey ideas but use language in STEM topics that, from a rhetorical perspective, attracts readers to be and become active participants of the story or research and not merely passive bystanders.

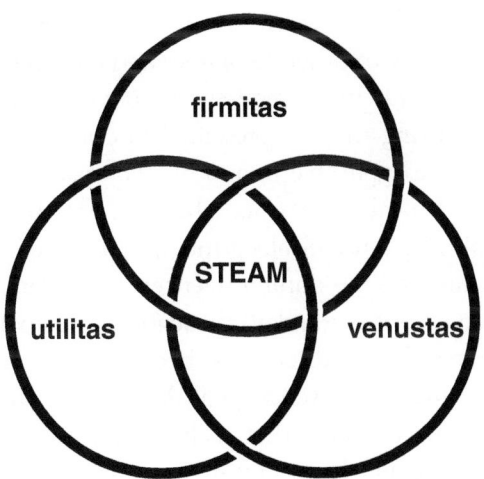

Figure I1. Vitruvian Trinity of Architecture Applied to STEAM and STEAM Writing

Overview of Book

Divided into three parts, this resource book will provide the STEAM reader with insight into how critical writing actually unfolds when writing in or about STEAM subjects. The first part of the book outlines the overarching framework of what it means to write in STEAM. Examples will be provided along with the similarities and differences when considering STEAM subjects. In addition, readers will be exposed to the different writing styles that scientists and STEAM practitioners and researchers use when writing in their prospective disciplines. We will examine how to write successfully in STEAM from brainstorming ideas to putting sentences and paragraphs into cohesive and organized structures.

The second part of the book discusses what it means to write in each of the STEAM subjects. Each of the four chapters in this part is comprised of at least three sections. Since writing, specifically critical writing, is intrinsically linked with reading skills (Farenga, et al., 2010), the first section of each of these chapters begins with approaches to reading and reviewing content. Since we will encounter the fundamentals of brainstorming prior to critical writing, the second section of each chapter in the second part of the book identifies the development of ideas in the STEAM discipline and puts those ideas into words. The third section provides STEAMers the opportunity to assess various examples.

In the third and final part, we investigate how to write in STEAM subjects as a form of integration with other disciplines and for practical purposes.

There are some overarching premises that traverse the entire book. First, certain themes in writing that are presented in one area of STEAM might, and oftentimes will, be applicable in another. For example, the development and structure of technical reports and the inherent rules within them for engineering can also be a useful blueprint (no pun intended) for those produced in architecture and the natural sciences. In a related manner, the use of mathematical equations is standard throughout the book, given their universality among the STEAM disciplines. Second, while I have adopted the style guide of the American Psychological Association (APA) for this book, STEAM writers should be aware that each STEAM discipline has either its own or crossover style guide. In this regard, I would strongly recommend each reader of this book to learn the style guide used in one's specific STEAM specialization. In Appendix D, I have provided specific examples of various ways to reference a publication in several style guides used in STEAM. Third, this book adheres to principles of social justice through critical pedagogy—a philosophical approach that encourages students to critique structures of political and corporate hegemony and subsequent oppression. Indeed, critical pedagogy is an exemplary approach for subject matter of this book, given that learning how to write in STEAM presupposes rigorous methods of data analysis, which require the student or professional to find the truth when answering a specific question. When we write in STEAM, our findings or outcomes are most often expected to create flux or disequilibrium in our investigations, especially if our alternative hypotheses turn out to be correct. Therefore, teachers will encourage their students to challenge existing long-held beliefs that often result in social and cultural inequities.

So, I welcome you, the reader, to join me on this writing voyage that will allow us to increase our understanding of written expression in STEAM in a way that motivates students, teachers, and professionals whose needs or interests span a wide array of STEAM subjects.

Daniel Ness
New York

· 1 ·
WHAT DOES IT MEAN TO WRITE CRITICALLY IN STEAM?

Perhaps you're good in the sciences, mathematics, or engineering, but you don't think you're good at writing. Or perhaps you're not a STEAM person but want to learn how to write in a STEAM subject. Perhaps you're a teacher who wants to improve your writing style. You might be a childhood educator and therefore want to write in a way that is clear and concise in any STEAM subject. Or you're a secondary school educator and you want to hone your writing skills in your specific STEAM subject. You might even be a college professor who specializes in a STEAM subject or who specializes in English, literacy, writing for teachers, or teacher educators who prepare teachers who teach English language learners (ELLs). This book is for anyone who belongs to any of these categories.

So many people specializing in one or more of the STEAM disciplines have one of the mindsets mentioned above—namely, that you don't think you're a good writer in your field, or, you're a good writer but need extra facility in writing in STEAM. But in fact, writing in STEAM is not all that much different than writing in any other academic or professional area—at least when it comes to ideas. Clearly, what is different is the content and mechanical syntax—special symbols and terminologies—within each of the STEAM disciplines. Writing in STEAM is not entirely like writing about Shakespeare, a topic in philosophy, or the history of music—although it can be depending on the circumstances of the topic. Writing in STEAM can, however, involve the use of equations, formulas, theorems, explanations of blueprints, and forms of written symbolism other than the standard alphanumeric characters.

But what about writing critically in STEAM? Are there any additional components to think about when engaged in the process of writing critically in STEAM? Before delving into critical writing in STEAM, we lead up to what critical writing is by thinking about what it requires the student or professional to know and do.

From Uttering Sound to Writing

While young children's tendency to begin engaging in oral communication is innate (through the guidance of universal grammar), their ability to read and write is not. The development from sound utterance to writing can be explained both from the perspective of the individual (ontogeny) and that of humans in general (phylogeny). From an ontogenetic perspective, language emerges when very young children begin to express themselves orally and comprehend oral language. They do this several years before they learn to use the written system of the primary language. It is important to stress at this point that many languages throughout the world are only oral and do not have writing systems. However, the opposite is not the case at all: no language has only a writing system without an oral language. But there's a paradox: for some written languages, including Aramaic and Latin, the spoken language has gradually ebbed over millennia and centuries, almost to the point of extinction. From a phylogenetic perspective, spoken language, in one form or another, has been in use for 50,000 years or more, but alphabetic (non-pictorial) writing systems are much more recent inventions, dating back to about 3500 B.C.E. Forerunners to alphabetic writing—letters on these pages that you are reading—are pictographic symbols, such as ideographs (shapes that represent concepts), cuneiforms (wedge-shaped symbols that represent sounds and meanings), hieroglyphics (pictures that denote objects, concepts, and sounds), and logograms such as Chinese, Japanese, and Korean characters that represent whole words or phrases. In short, reading and writing are very recent human communicative activities from an evolutionary perspective.

In parallel, reading and writing are learned activities that will take years for young children to hone and polish. The basic gap between the spoken and written word is the symbol—its visual representation and its meaning. The semiotic function is what Jean Piaget calls the ability of the very young child (usually at toddler age, or post-infancy) to make the connection between symbol and concept. The teacher's knowledge of semiotics, the study of the properties of signs and signaling systems found in all forms of written communication, is important when considering that children and students of all ages begin at one point or another to acquire symbols for representing concepts. After learning to read, children begin to write by learning the formation of letters, learning the spelling of speech sounds, developing skills in handwriting legibly, and beginning to find ways to share their ideas by finding the best ways to convey meaning through written communication. Learning how to write, let alone write well, requires the individual to develop an understanding of the conventions of sentence and paragraph structure, grammar, usage, mechanics (such as punctuation, capitalization), and organization of longer pieces of writing. As students develop their ability to write, they will learn the different genres for writing (essay, poem, technical report, letter, research paper, book, journal article, etc.) and each of their special requirements. Eventually, they will learn about the processes of writing, namely, brainstorming, outlining, drafting, revising, and editing (as discussed in Chapter 3).

As children develop, they enter and leave different stages of reading and writing development. But these stages are not innate or automatic; nor are they acquired through behavioral conditioning. Rather, they are grounded in cognitive development through direct or experiential learning. Children, adolescents, and both native and non-native English-speaking older students and adults require instruction and practice. Moreover, not everyone learns to read and write in the same way or by following the same instructional sequence. This is due to many variables, which include, but are not limited to the following: teachers' and instructors' variation of the procedures they use to develop literacy; the diversity of students' knowledge and abilities in the same class; the teachers' own writing experience, knowledge, and even pedagogical philosophies; the availability of time and resources; and the directives of administrators, policy makers, and those responsible for standards that teachers are compelled to follow.

Critical Literacy in STEAM

We now arrive at the development of *critical literacy*, which is the ability to recognize and realize power relationships and class struggle in society. Situated within the area of critical pedagogy, critical literacy is an approach in which writers wrestle with topics by addressing and challenging oppressive ideologies, discourses, and practices in order to enact social change (Janks, 2013; Luke, 2012). Critical literacy specialists critique written materials that directly or indirectly support class oppression. Critical literacy is a consequential area of study that emanated from critical pedagogy, a teaching paradigm introduced by Paulo Freire (1970). Using the term "cultural conscientization," which derives from the concept of "concientização," Freire refers to the ways in which we learn about socioeconomic and political contradictions for the purpose of challenging the disenfranchisement of marginalized populations. Like Freire, critical literacy specialists argue that literacy is not neutral, and that reading and writing are shaped by social processes and patterns of power within social settings (McLaren & da Silva, 2001).

With respect to STEAM subjects, Knijnik (1993), following in the footsteps of Freire, outlined and developed methods and mathematics curricula that members of subordinated cultural groups can possibly implement as a means for emancipation and perhaps liberation from governing bodies that enforce hegemonic political structures. One of these methods, for example, concerned the learning of reading and writing mathematically as a channel for opportunity and awareness. In particular, she investigated oral mathematical knowledge as a form of social and political mobility from the vantage point of landless Brazilians. Knijnik transferred this mathematical knowledge to a system that enables teachers to redirect the oral and written mathematical tradition so that landless people can appreciate how mathematics can be used for advancement and liberation. This Freirean perspective, which has empowered landless Brazilians through reconsidering how mathematical knowledge is learned, counters the status quo set by governmental institutions and policymakers that has traditionally ensured a permanent underclass of mostly minority, minimum wage workers. In institutionalizing the status quo, schools have paid virtually all their attention to the memorization of facts and little, if anything to the development of critical literacy.

There has been no better time to apply critical literacy to STEAM than the present. Given widespread misinformation, disinformation, and malinformation in the popular media, the scientific community at large has gone through exasperating periods of unwarranted scrutiny by several politicians and lawmakers who have attempted to downplay scientific evidence in favor of political advantage or control. An aspect of fringe thinking during most times in history, this unjustified inquiry is based on no scientific evidence, yet it is unflinchingly supported by uninformed or apathetic populations. In countering these efforts, critical literacy has recently been emerging as a central theme—not only in the social sciences, but in the natural sciences as well.

From gleaning the research literature on scientific thinking, scientific learning, and critical literacy, Krajcik and Sutherland (2010) identified five instructional and curricular qualities that can support students in developing their critical literacies in the context of science. These qualities are as follows: (1) linking new ideas to prior knowledge and experiences, (2) anchoring learning in questions that are meaningful in the lives of students, (3) connecting multiple representations, (4) providing opportunities for students to use science ideas, and (5) supporting students' engagement with the discourses of science. The purpose of these five qualities is to promote students' ability to read, write, and communicate critically about science so that they can engage in inquiry with others in their lives. In adopting Krajcik and Sutherland's approach, I would argue that this model can be extended to serve all STEAM disciplines, not only those in the natural sciences.

To begin, thinking and writing about STEAM critically would require the individual to embrace a constructivist, feminist, queer, neo-Marxist, critical race, or human developmentalist philosophy. Any other mostly reactionary philosophical or theoretical foundation would essentially contradict or upend the goals behind critical literacy and critical pedagogy. To this end, the ability to link new ideas to prior knowledge and experiences undergirds our ability to weigh evidence from data and make conclusions from that evidence. Prior knowledge comes from at least two sources: past experiences and school content that was learned on an earlier occasion. So, eliciting prior knowledge is critical when encountering abstractions in our daily lives. Let's take epidemiology, an important area in STEAM, as an example. Students throughout Kindergarten and into high school will undoubtedly catch a common cold or possibly another type of viral infection at some point during their childhood or they will encounter a sibling or parent who comes down with this condition. Children and adolescents, as well as many adults, will often remark that a common cold is a result of staying out in the cold weather for too long, being overworked, or not getting enough sleep. But these are only tangential reasons for catching a viral or bacterial illness. Because of the highly abstract nature of viruses, Krajcik and Sutherland point out that school science curricula often fail to connect students' prior experiences and prior knowledge, a situation which often leads to a lack of interest and poor performance in middle school and high school science courses. Children and adolescents do know, however, that other people who are in close physical proximity to someone who comes down with a viral illness are susceptible themselves to becoming infected with the same virus and will often end up being unwell. When teachers and the predominantly prescribed science curriculum do not address the connections between viruses and other submicroscopic agents to physical illness and prior knowledge, students will lack the foundational underpinnings of life science—from the submicroscopic, miniscule agents and organisms to observable life forms. Clearly, these newly infected people did not become infected from cold weather or working too hard. Accordingly, students need to be aware of cause-and-effect relationships in their prior experiences. The same is true for each of the other STEAM disciplines. In mathematics for example, students who play musical instruments almost always understand the need to play notes in different time durations—some have as long as the first note, a quarter as long, a fifth as long, twice as long, and so forth. But again, prescribed curricula often fail to make the links between everyday concepts and in-school, scientific concepts explicit, thus impeding students' conceptual knowledge. The idea of connecting students' prior experiences and knowledge with in-school concepts is not new and is a product of the works of many philosophers, educators, and psychological theorists—from Jean-Jacques Rousseau to Maria Montessori, Jean Piaget, and Lev Vygotsky. In Vygotsky's own words:

> In working its ... way upward, an everyday concept clears a path for the [in school] concept and its downward development ... [In-school] concepts, in turn, [allow] for the upward development of the child's spontaneous concepts toward consciousness and deliberate use. (1934/1986, p. 194)

The second quality, anchoring learning by initiating interest and motivation through meaningful questions that affect the lives of students, can also be applied to all STEAM disciplines. For example, let's consider the following two questions: 1) In C++ programming language, explain the meaning of the objects **std::cout** and **std::cin**. 2) How would you write a program in C++ for developing a Graphical User Interface (GUI) application for architects to prepare blueprints? If we compare the two questions, the first one asks the computer science (i.e., technology) student to provide an answer that requires low-level knowledge. According to Bloom's Taxonomy of Educational Objectives, the student is asked to simply explain something based on definition and possibly description. That is, it's not the type of question that will generate high levels of motivational and cognitive currency. On the other hand, the second question requires a higher cognitive workload and is a question that invites the student to embark on a motivating journey; it's a project that would yield rewards because the results or outcome generate several affordances for the user of the program. So, the second question is clearly the more effective in terms of both motivation and cognitive development.

The third quality, integrating text and visual representations, is an important part of STEAM critical literacy. Picture a science class in which the teacher presents the DNA molecule solely through lecture and textbook reading. In order to advance critical literacy in science, the teacher can present a supersized DNA molecule model for the class. Better yet, the teacher can ask students to create their own models of the DNA molecule. In mathematics class, rather than teaching the irrational number π by merely plugging it into an equation for solving, have students take a hockey puck to determine how to derive π by taking specific measurements of the circle. The basic idea here is that STEAM critical literacy can often be developed through a combination of written communication and visual representation, and students' prior experiences.

Providing students with time, opportunities, and guidance to apply science learning to new contexts, the fourth quality, is also adoptable in all STEAM subjects. For example, how would a student of architecture write about multiplication—a mathematics subject—in the context of dimensions on a blueprint? Or a physics student can take the theoretical foundations of motion and apply it to spacecraft by writing about rocket propulsion. In another instance, a music theorist can apply the physics concepts embedded in acoustics and sound to develop a novel way to divide individual pitch intervals into sevenths, thus producing a new scale for microtonal music performances. As another example, everyday knowledge provides us with the understanding that some filled balloons keep going up into the sky while others don't. The elementary or middle school student would benefit by learning about different gaseous elements and which one will lift the balloon and which ones will not.

Fifth and finally, Krajcik and Sutherland (2010) discuss the importance of engaging in the discourses of science—that is, "... explicitly supporting scientific discourses, including the language of science and its practices ..." (p. 457). In line with critical literacy, they go on to argue that students must "have the opportunities to talk and write about science and to practice supporting their ideas with evidence" (p. 458). I cannot agree more with Krajcik and Sutherland on this point. In fact, we again can apply the importance of engaging students in critical discourse, not just in science, but in mathematics, technology, engineering, and arts as well. In fact, the overarching theme of writing critically in STEAM embraces the importance of critical dialogue as it relates to written communication for the purposes of challenging commonly held perspectives and inviting readers to reconsider newer possibilities that can potentially be more effective than traditional ways of thinking and knowing.

Differences and Similarities Among STEAM Subjects

Now that we've been able to consider some of the generalities of writing critically in STEAM, we can embark on a discussion of differences and similarities among STEAM subjects. In this section, we will begin by identifying key differences. Knowledge of differences in approach to writing among the STEAM subjects is important because it allows you, the expert, dilettante, student, or teacher, to focus on the tools that are needed in your particular STEAM area of study. Note that our discussion here will be generalizable across each of the STEAM areas. For more in-depth coverage of writing in one of the STEAM disciplines, I would refer the reader to Chapters 4, 5, 6, and 7—the second part of this text, which are exhaustive and detailed accounts of how to write critically in each of the STEAM areas.

Based on the STEAM writing examples that we will analyze below, it should be clear that writing in one STEAM discipline is not necessarily the same as writing in another. For example, while there may be overlap in terms of content, writing in one of the natural sciences is not necessarily the same as writing in mathematics. In the next chapter, we will examine STEAM writing styles more closely, and look at the scientific delivery and spontaneous delivery styles that help writers in STEAM make their points clearly. In terms of writing in academic disciplines, writing in STEAM differs from that of non-STEAM disciplines in at least four ways:

1. Special symbols and terminologies
2. Structure and form
3. Research and analysis
4. Findings and conclusions

Special Symbols and Terminologies

Perhaps the most glaring difference that you will encounter when reading professional, layperson, or student writing within and among the STEAM disciplines and non-STEAM disciplines is the use of different symbolic systems and terms. As we shall see in subsequent chapters, there will be symbols that you've experienced and worked with and others that you may not have seen before. For the non-STEAM person, with some exceptions, many of these symbols will be entirely unfamiliar to them. Now it is true that many science, engineering, and technology professionals will have a fairly wide-ranging mathematical knowledge. But some areas of mathematics that are useful for the engineer might not necessarily be useful for the chemist and the economist (Thomson, 2001)—believe it or not, another mathematics person—who will have uses for mathematics that are different from those used by the psychologist—again, believe it or not, successful psychologists know their math!

We discussed semiotics earlier in the chapter, and we said that semiotics is the study of signs and symbols. We also said that, from an ontological perspective, the semiotic function refers to the very young child's emergent ability to make associations between ideas and words. Keep in mind, though, that "words" in this case are the sounds of words and not the notion of a sequence of adjacent written letters of the alphabet that are separated on both the right and left sides by spaces (as you see in the sentences and paragraphs written here). For example, infants under the age of 1 year are able to produce sounds that emulate what a parent or an older sibling might say, but this is not evidence of the semiotic function because the infant is only mimicking a sound or a word that they might hear someone else say. However, by the time the very young child reaches a slightly older age, say, about the end of the 2nd year, there is usually evidence that the child will be able to connect the word with the thing or idea that the word represents. In linguistics, we say that the word is the signifier, or the intermediary, and the thing or idea the word represents, is the signified. So, when a child associates the word "bird" with a creature that the young child sees flying in the air, "bird" is the signifier and "the flying creature" is the signified—the actual thing or idea. The same idea holds true when we are older and learn new symbols in different academic fields. The Kindergartener or first grader connects the "+" with combining things together. We learn that "—" means to remove a number of things from a larger number of things in elementary school, but for the engineering student, "—" refers to the straightness of an object. In mathematics, we learn that $\emptyset$ is the empty set in middle school, but again, for the engineering student, $\emptyset$ means diameter of a circular object. The symbol Δ in the physical sciences means "change," but it has entirely different connotations for the elementary school student, who will most likely interpret it as a triangle. And of course, the $\int$ symbol means integration and tells the mathematician to find the area under a two-dimensional curve, while integration allows the civil engineer to find the area under a curve in order to maximize an object's function and minimize the same object's potential of failing, bending, buckling, or collapsing. The basic idea here, then, is to emphasize the importance of symbol inclusion in STEAM writing and that each discipline requires a different set of nomenclatures that serve as signifiers. The novelist's signifiers are special uses and variations of words. The STEAM student or professional needs to integrate words with other symbols. So, when an engineer is testing the strength of the Golden Gate Bridge in San Francisco, by writing "$\int_{-640}^{640}\sqrt{1+\left(\dfrac{640}{663}\dfrac{e^{x/1326}-e^{-z/1326}}{2}\right)^2}\,dx = 1326.956$," the idea is to determine the extent to which the suspension bridge will not buckle under stress due to different levels of earthquake readings. In this case, "$\int_{-640}^{640}\sqrt{1+\left(\dfrac{640}{663}\dfrac{e^{x/1326}-e^{-z/1326}}{2}\right)^2}\,dx = 1326.956$" is the signifier, and the actual bridge itself is the signified.

Structure and Form

As you embark on writing in one of the STEAM areas, you will also notice that the structure and form of the letter, research paper, article, book, or digital communication differs depending upon which STEAM subject is considered. You now might be thinking what I mean by structure and form. In structure, I'm referring to the overall shape, appearance, layout, and configuration of ideas and content. In form, I'm indicating the overall procedure or formulaic practices that each of the STEAM disciplines use to write in their field. For instance, an engineering research article might possess the same structure as a research article in biology, but the two disciplines will more than likely differ in terms of nomenclature, which will therefore affect form. I discuss structure and the different forms in each of the STEAM specific chapters, namely, Chapters 4 through 7.

Research and Analysis

While engaged in the writing process, specifically if you are a professional or a student, you will also notice differences in terms of your plan of research and analysis. Research encompasses a wide array of analysis in different academic fields. The first thing that many of us associate with scientists is that they conduct experiments, even if we might not be sure of what an experiment is supposed to do and tell us. Experimentation is the method one uses to identify the existence of a cause-and-effect relationship between two things. I tell my mathematics and mathematics education students that experimentation is also a method that mathematicians use in their research, obviously depending on the kind of mathematics one is studying. It should be noted that while experimentation is frequently used outside of traditional STEM subjects, such as in psychological and sociological research, it is a method that has its origins in STEM disciplines and has been part of STEM research ever since. Regardless of discipline, an experimental design is the process that involves steps to follow so that the experiment can be conducted. Moreover, these steps must be completed in sequence so that the experiment is carried out correctly. In general, the experimental design is generally written and outlined in a section called "Methods."

In contrast to STEM designs, there are certain research designs, which tend to work better in non-STEM subject areas. For example, textual criticism is a method commonly used by literary scholars and theorists for the purpose of identifying meaning of text through interpretation and comparison with other texts. Biblical historians and scholars have been successful with this approach in deciphering meaning of written passages through comparison and historical context. Other examples include naturalistic observation, semi-structured interviews, and case studies, all of which are common as research approaches in the social sciences.

Findings and Conclusions

Out of all the sections of a research paper in STEAM disciplines, perhaps the discussion of the findings is the place that most people interested in your study will turn to. This is because readers want to know the essential upshot of what was found and how the findings reflect present and future investigation. It is at this point when the writer reveals the research findings in light of the original hypothesis. You can now address the following questions in the text: What was the purpose of my study? How do my findings relate or connect to that purpose? To what extent were my findings of interest due to chance? How valid and generalizable were my findings? And are there any larger implications based on my findings?

In addition, the STEAM writer will include the benefits of the research findings and how these benefits will impact both the science community and possibly the community-at-large. If serendipity or chance played a role, this is the place in the paper where the writer details the unexpected by-products and outcomes. It is also in the discussion when the writer becomes dissenter and critic and provides an overview of the paper's shortcomings or inconsistencies, as well as implications for future research.

Conclusion

In this chapter, we showed that writing is not an innate ability; nor is it a skill that involves conditioning techniques, such as stimulus-response, that is commonly referred to as behaviorist techniques. Rather, writing, whether STEAM-related or not, is a learned process that involves one's access to prior knowledge and experiences as well as practice in shaping the specified conventional symbols and lexicon that are available to us. We have also seen how writing in STEAM can and must be conducted with a critical and impartial eye. Lastly, we have identified some of the common similarities and differences of STEAM writing and non-STEAM writing. In the next two chapters, we will address the issues of tone, voice, verb tense, and usage errors in STEAM.

· 2 ·

TOOLS FOR WRITING CRITICALLY IN STEAM

Now that we have some of the basic ideas of critical STEAM writing in place, we're ready to start acquiring the necessary tools to learn how to write critically in STEAM. Before we begin, though, it will be necessary to outline our plan of action. First and foremost, as writers in any subject, not only those in STEAM, we need to address individuals with whom, to whom, or about whom we are writing with respect and dignity. That is, as writers, we need to be critically conscious of our focus of content and the people who might be the subjects of our central theses. Freire's concept of cultural conscientization, or critical consciousness, that we have introduced in the previous chapter, underscores the importance of writing about issues in ways that indict sexist, racist, homophobic, xenophobic, and generally regressive and reactionary mandates and points of view.

As a post-Marxist concept, conscientization provides the writer with the tools necessary to address commonly held positions of misinformation, disinformation, and malinformation that have historically marginalized and oppressed specific groups. From a STEAM perspective, epidemiological data provide a fruitful example of how written and verbal communication over the years have been used as fodder for all three types of perversion. Misinformation consists of falsified accounts, stories, or reports that have either not been challenged or scrutinized or those that are shared by individuals who unknowingly believe that the information is truthful. For example, social media bloggers who propose alleged "treatments" to viral infections spread misinformation if they are unaware of the scientifically proven and successful results of vaccines. Unlike misinformation, disinformation occurs when an individual knowingly fabricates information for social or political gain. For example, a politician or pundit who writes an editorial in a newspaper blaming an opposition party for the spread of a disease is circulating disinformation. As another example, a reputable scientist who is rebuked by a politician posing as a physician who posts erroneously damning propaganda on their website for profit is disseminating disinformation. In contrast to misinformation and disinformation, malinformation, which can be just as nefarious as disinformation, is a situation in which a person or group uses an actual datum to demonize and scapegoat another, oftentimes marginalized, population. Using the pandemic example, malinformation occurs when an individual or group of individuals intentionally and calculatingly excoriate the ethnic or cultural group whose ancestries emanate from the purported region or country of origin of a particular virus or illness.

The second part of this chapter examines important issues in grammar and mechanics that the writer in STEAM should employ. Like issues of critical consciousness, learning essential tools of grammar and mechanics is a process that all writers, not solely those engaged in STEAM, will need to consider. We start by considering the importance of tone—the way in which your writing is received by the reader. We will then move to the suggested use of voice. Next, we consider verb tenses and subject-verb agreement. The chapter will close with a discussion of common usage errors, appropriate applications of punctuation, and the suitable uses of quotations.

Critical STEAM Writing and Gender

Any document, paper, or publication must be written in a way that accounts for gender equity. Considering gender equity not only sustains and supports a just society but also leads the author to write with precision in mind. Therefore, when referring to females, males, and people who identify as non-binary, the writer in STEAM should avoid the antiquated use of the masculine gender in nouns and pronouns. For example, rather than writing, "He who is an inquisitive scientist believes that his laboratory is his playroom," the author should instead write "Inquisitive scientists believe that their laboratories are their playrooms." Notice the switch to plural form when rewriting this sentence. As another example, write "People are social animals" rather than "Man is a social animal." The plural alternative, then, is an appropriate, yet unique, way to remedy the problem of gender inequity.

While this solves the problem of gender generalization, it does not necessarily address the issue of addressing individuals who identify as transgender. Because one's identification must be valued and respected by all communicators, the writer must realize how the term and meaning of *transgender* is to be used properly in any form of communication, not solely written correspondence. Transgender is a broad term that one uses to describe someone whose gender identity is different from that person's sex assigned at birth. Gender non-binary is a related term referring to a person's gender identity that does not conform to traditional female or male categories. So, there are three singular, pronominal alternatives, not two: 1) she/her, he/him, and they, their. This topic is explained further in Chapter 6, when we encounter email writing and the importance of salutation in formal written letters.

Tone

STEAM writing, like writing in other academic or professional domains, requires the writer to use the appropriate tone for the audience to whom the written text is directed. Tone is discussed at greater length in Chapter 6, when we examine tone in terms of the approach that engineers or architects use when writing letters or related forms of correspondence to clients or colleagues. For now, though, it will be important to identify some key criteria to consider for STEAM writers in general.

Direct vs. Indirect Introduction

As noted earlier, while many of the ideas herein are directed to burgeoning writers of any domain, beginning and experienced STEAM writers are frequently noted for thinking that readers will be uninterested in what they have to say. One of the overarching goals that has been emphasized throughout this book, then, is to limit the possibility of this lack of interest or curiosity. One way of doing this is to consider how written text should attempt to maximize the reader's attention. Writers have used two possible rhetorical approaches to draw readers' attention: Direct and Indirect Introductions. Instructors often ask students to begin their essays or writing projects using a direct approach. The rationale for doing so is clear: Tell the reader up front what the purpose of the paper is and how the issues within it will be addressed. This is a valid suggestion, especially when writing on a STEAM topic; when writing up studies in the natural or social sciences, we're often asked to write in short, pithy prose. There are, however, drawbacks to the direct approach. First, direct introductions can often be construed as harsh and laconic in tone to the reader. For certain colleagues, especially those in a superior position, the direct approach, as a form of initial delivery, can seem condescending. Another drawback to the direct introduction, also related to the tendency toward a terse writing style, has to do with the potentiality of boring the reader; this approach may be uninspiring and limit the reader's interest or stimulation. Therefore, this might not be the right approach to consider when writing books and for certain—indeed not all—academic journals. The following is an example of the introduction to a short article from a practitioner-based journal (Ness, 2005, p. 59). Notice how it begins with the purpose of the article and the subsequent sentences expound on the first:

> In this [article], we experience, through simulation, the skills that are required of a cartographer. A cartographer is a person who practices the art of making maps. Cartographers' maps were often unique, visual representations of data. When examining different maps that were created in different centuries, one notices how various geographic features of the Earth, such as bodies of water, mountains, and deserts, have been depicted in a variety of ways. In examining the maps, it is evident that the style of the map was affected by the mapmaker's perspective of the world.

An indirect introduction is one in which the writer introduces the main idea after the first paragraph or section of the paper. It often begins with a paradox, which enables the writer to convince the reader of the importance and novelty of the author's argument or central thesis. This paradoxical situation is important because it provides the writer with the liberty to demonstrate stark contrast at the outset so that the reader has the opportunity to consider the point of view that is contrary to popular or current belief. An indirect introduction can also involve the use of analogy that provides the initial backdrop of the writer's central thesis. The following excerpt is an example of an article with an indirect introduction (Farenga, Joyce, & Ness, 2004, p. 60).

> It is often said that a dog is your best friend. In fact, after cattle, dogs are the oldest domesticated animals. Dogs have lived with humans for over 10,000 years and have been selectively bred for various types of domestication. Domesticated dogs have performed many roles in society throughout history—rescue dogs, hunting dogs, guide dogs, and police dogs to name a few. Beside dogs, humans have opened their homes to numerous organisms, including cats, birds, lizards, snakes, and rodents. Yet, how much do we really know about the animals that live with us? And, have these animals learned from us?
>
> Sometimes we know more about animals in the wild from research journals, television documentaries, and science researchers, than the animals in our own homes. The study of animal behavior is called ethology. Ethologists have provided systematic observations over time to highlight complex social relationships of primates. You can foster scientific observation by having your students conduct similar observations with animals at home or at school.

Voice and Critical STEAM Writing

Next, we consider voice and its importance in STEAM writing. The forms of the verb used in writing can be expressed with one of two voices—active or passive. One writes in the active voice when attempting to represent the subject of the sentence as performing the action expressed by the verb: *We conducted the experiment.* One writes in the passive voice when the subject of the sentence undergoes the action expressed by the verb: *The experiment was conducted by us.*

Let's metaphorically cut to the chase: The active voice is more direct and compelling for the reader than the passive voice is. Take the following example, and consider which sentence, either when written or heard, seems to be more convincing to the ear:

> **Active voice:** Pursell demonstrated an example of gender disparity in marketing science toys.

> **Passive voice:** An example of gender disparity in marketing science toys was demonstrated by Pursell.

For most readers and listeners, the active voice example is leaner and more explicit than its passive counterpart. In colloquial terms, the active voice example tells it like it is, whereas the passive voice example seems indirect and circuitous, more cerebral, ornamented, and therefore not as straightforward as the active voice. However, there are cases in which the passive voice should be used. One such case has to do with our previous topic—gender. Passive voice can at times be more useful when attempting to refer to all genders and avoiding gender traps of singular masculine pronouns. Take the following example, which is an attempt to generalize about all mathematicians:

> **Active voice:** The mathematician knows his axioms.
> **Passive voice:** Axioms are known by the mathematician.

In this case, the passive voice example circumvents gender traps, which would otherwise result in constraining the writer to use a singular gender pronoun, most often, "he" or "his."

Verb Tense

Selecting verb tense when writing is an important consideration. For one thing, writers should be consistent in what they are attempting to say. I have seen papers that are present tense in one sentence and past tense in the next. The following rules of thumb are useful in writing a STEAM research paper of any kind.

1. In reviews of literature, use past tense to report studies that were conducted in the past. For example, "McBride (2021) *pointed out* that . . ."
2. Use past tense when writing the Method and Results sections because the very fact that these sections are being written assumes that these parts of the study were already conducted (i.e., the past). For example: "Surveys *were distributed* to 45 students . . ." or "Services rendered on the structure *were completed* on . . . [sometime in the past]"
3. Use the present tense to define terms: "Astrophysicists *define* 'Goldilocks exoplanets' as those that may support life."
4. One can also use the present tense when stating a hypothesis or a claim: "According to Marie Curie, radioactivity *is* a state whereby an atom's nucleus *becomes* unstable and loses energy due to radiation."

5. In terms of research papers, future tense is used almost exclusively for the Discussion section, when the author identifies how society can benefit from the results: "Female students *will increase* their motivation in STEM if this initiative is implemented." Future tense allows the author to introduce implications from the findings that need further investigation or analysis: "Future research *will be* necessary in order to . . ." But note also that this statement can also be written in the present tense: "Future research *is* necessary in order to . . ."

Usage Errors in STEAM

Usage errors are common in STEAM letters and related forms of communication, reports, or research papers. Most common errors in usage arise when writers use the incorrect homonym, namely, the spelling of a word that sounds like another word that is spelled differently. Three common STEAM homonyms that are often confused are 1) "principal" and "principle," 2) "complement" and "compliment," and 3) "affect" and "effect." Usage issues also arise when considering the singular and plural forms of some familiar STEM-related terms:

Singular	*Plural*
Alga	Algae
Analysis	Analyses
Anomaly	Anomalies
Appendix	Appendices
Cortex	Cortices
Criterion	Criteria
Datum	Data
Deer	Deer
Focus	Foci
Fungus	Fungi
Hypothesis	Hypotheses
Larva	larvae
Matrix	Matrices
Ox	Oxen
Phenomenon	Phenomena
Serum	Sera
Species	Species
Stylus	Styli
Virus	Viruses

The following are additional usage errors in STEAM papers. These errors have to do with specific forms of Greek and Latin prefixes and roots as well as Greek and Latin words that eventually became Anglicized. To begin, one major error in STEAM writing in particular results from the confusion of *data* with *datum*. The word "data" is plural and "datum" is singular. The problem is that most writers wrongly believe that the word "data" works for both singular and plural and almost always mistakenly use the singular verb with "data": "The data shows . . ." is incorrect. The correct forms of this term are: "The data show . . ." or "The datum shows . . ." Another common problem occurs when considering the prepositions "between" and "among." The key to understanding the meaning of these two words are as follows: Use between when referring to two terms only. Use among when referring to more than two terms.

With these correctable errors aside, there are situations when a term may be used incorrectly but there is nothing one can do as a writer to fix the correction. One situation in mind occurs in statistics—particularly when running an analysis of variance (ANOVA). Common parlance when conducting an analysis of variance is to refer to the "between the sum of squares." This statement is still used when the number of conditions being compared is greater than two.

Common prefixes in STEAM writing that are frequently confused include differences between inter- and intra-, intro- and extra-, hypo- and hyper, and ante- and anti-. "Inter" means between while "intra" means within. For example, interpersonal relationships occur between or among two or more people while an intrapersonal temperament is the psychological disposition of an individual. An introvert and an extravert are two people with diametrically opposite dispositions; the introvert, Clark Kent, keeps to his own while Superman, Kent's alter ego, is an extravert who is not afraid to say or do anything. Hyper- means excessive while hypo- means deficient. So, hypertension is excessive tension or blood pressure that is above normal while hypotension is the lack of tension or blood pressure that is below normal. Finally, ante means before while anti means against or not in favor of. Antebrachial symptoms are health patterns related to the forearm while an anticoagulant is a medicine that prevents blood clots. Please consult Appendix A, "Commonly Confused Words" for a large selection of commonly confused words.

STEAM Writing Examples

The following paragraph is an example of what it means to write in statistics—a branch of mathematics. Notice possible similarities and differences

TOOLS FOR WRITING CRITICALLY IN STEAM   27

between this excerpt and other passages from a history book, an English language primer, or selections from a newspaper or magazine article.

> Statistics have to do with the idea of processing numbers in such a manner in order to produce clear, concise, and representative information that is readily consumable and ready for use. One particular statistic that you are probably already familiar with is the average. Suppose you wanted to know the average age of students in a particular classroom. The process of obtaining the average age of students in a classroom is relatively simple: 1) ask each student his or her age and have each student write his or her age on a slip of paper; 2) then, collect the slips of paper; 3) add up all the ages on each of the slips of paper; 4) lastly, divide the sum of the ages that you obtained in the last step by the total number of students in the classroom. A sample case might be as follows: You obtained the following ages and you are now going to add them up. The ages are 12, 14, 12, 13, 13, 13, 13, 15, 14, 12, 14, 12, 13, 12, 12, 13, 14, 12, 13, 14, 15, 13. You arrived at the sum of 288. Then divide 288, the sum by the total number of values, namely, 22. The answer you get is 13.09. This means that the average age of the students in this particular class is 13.09 years.

One of the most important features of writing in STEAM when it comes to writing in general is the idea that the grammar and syntax is universal; we don't change the grammar to write in STEAM or in any other domain for that matter. So, take as an example the phrase "ask each student his or her age and have each student write his or her age on a slip of paper . . ." It should be noted that the adjective "each" is singular, and therefore takes on singular possessive pronouns. So, it's good to know that grammar is unchanged no matter what domain you plan to write in. But what's different? What is different when it comes to writing in STEAM when compared to writing in other, non-STEAM disciplines? In the above example, we have numbers—lots of them. A whole row of 22 numbers, some of them repeating. The point of these numbers is not whether they repeat or not, but the fact that their listing is random and based on the ages, each of which was placed on a slip of paper. What else seems to be new here? This example demonstrates the use of a mathematical equation. Now the equation is not written out but it doesn't have to be. Rather, the equation is an explanation of how to obtain the mean or average of 22 numbers—namely, add the 22 numbers together and divide by 22 to obtain the mean. We could have written out the following:

$$\frac{288}{22} = 13.09 \quad \text{or} \quad \frac{\text{sum of the values}}{\text{number of values}} = \text{average}$$

But it is clearly all right to explain this equation in prose—that is, in a written narrative that does not show the computational equation or formula.

The following passage (Ness, 2022, p. 17–18) is an example of what it means to write in civil engineering—a branch of engineering that has to do with the structures that we see around us: roads, bridges, buildings, skyscrapers, and the like:

> Compression and Tension: Beams, columns, and cantilevers are structural elements. But tension and compression are physical properties. Structures cannot remain standing without the dynamic processes of compression and tension. Compression is an internal force that causes a structural element to shorten. This is precisely what beams do to columns—they shorten them. This might not be able to be seen with our eyes when viewing the inner workings of a building or bridge in the process of construction. Nor can it be seen in the block constructions of children. But compression is present; architects and engineers can tell you that buildings, bridges, and towers undergo compression-tension subtleties; without them, these structures would not remain standing due to brittleness, or the lack of ductility—the tendency of a material to fail suddenly and catastrophically, without plastic (i.e., stretching) deformation. Problems arise, however, when beams become too heavy and columns too thin or unsteady, in which case the column will buckle. Both professional architects and engineers and emergent architects and engineers as young children engage in behaviors that involve structural elements to be in a state of compression, as seen in the examples above.
>
> Tension, in contrast to compression, is an internal force that causes a structural element to elongate or stretch. Tension is what happens to beams, which are supported at two ends by columns or other elements that keep beams from falling. In addition, live loads add additional stress or tension to the tension member, which could be a beam, deck, or cable. Tension members undergo elasticity that prevents them from snapping or breaking—assuming there is no strain on the members. Similar to compression, it is important to note that tension is a process that may not necessarily be seen, but it is present. Again, we see tension in action in both examples of professional architects and engineers and those of emergent architects and engineers as young children.

What is different in this engineering example from what we found in the mathematics example earlier on statistics? For one, there are no numbers or equations per se. This is not to say that engineering texts don't have numbers and equations in them. Many do. But this one focuses more on the idea of terms and their meanings. For example, the author of the passage defines compression as "an internal force that causes a structural element to shorten." The author goes on to define tension as "an internal force that causes a structural element to elongate or stretch." Clearly, these are topics in engineering because tension and compression are key forces that are essential in ensuring that a bridge does not fall and that a building or skyscraper remains erect.

Getting Back to the Issue of Writing Critically in STEAM

Let's get back to the topic of writing critically in STEAM. To write critically in STEAM, first and foremost, means to write critically within the domains of STEAM. Writing critically means that we need to consider how influential and powerful writing is in general and how it can be used to the advantage of students and professionals alike. Students in STEAM-related disciplines should feel energized and encouraged to write in their fields of inquiry without hesitation or disinclination. STEAM practitioners and professionals should feel empowered as teachers who can serve their students and society at large with up-to-date strategies in helping their students and the public write in STEAM-related disciplines. Onward—let's get down to the essentials of writing in STEAM and what it entails. We begin with the pre-writing process of brainstorming and continue along the writing continuum toward the drafting stage of your STEAM writing composition in the next chapter.

Chapter Activity

In the following paragraph, identify as many errors as you can. Note that these errors come in many forms and include problems in both grammar and mechanics.

> As children's mathematical thinking develop, they begin to economize their counting strategies. The disadvantages of pushing aside is that it takes to long to count the total. A more sophisticated method, then, is tagging. The strategy of tagging involves touching or pointing to each object as they are being counted. When compared by pushing aside, children may be more efficient when counting by tagging. And tagging can get cumbersome and inefficient, it still requires tallying each member of a set. As a means of acquiring counting efficiency, subitizing, in which they show immediate recognition of a set's cardinality, is creatively employed by children. Subitizing, a word borrowed from Italian ("subito"), which means "immediately" or "spontaneously," involve a child's eventual ability to recognize the total amount of objects in a group fairly quickly. The child's ability to subitize mark the starting point to the more advanced forms of counting members of both sets mentioned earlier—namely, his ability to "count on," rather than "count all." Based on clinical interview

data, which demonstrates children's developing ability to instantaneously recognize various amounts of objects, subitizing defines the emergence of the edition concept. Accept for a few instances, children have been shown to excel in their adding and subtracting skills.

· 3 ·

FROM BRAINSTORMING TO WRITING CRITICALLY IN STEAM

Most people who practice their writing skills in any academic or professional domain succeed by brainstorming prior to producing a manuscript. Even experienced writers need to think through what they want to write before they put pen to paper or fingers to keyboard. However, rookie and veteran writers do not always brainstorm the same way. The first part of this chapter, then, is to identify reliable strategies for brainstorming ideas when about to write in STEAM-related subjects. The second part of this chapter focuses on subsequent steps that occur after brainstorming takes place. These steps are based on identifying ways to categorize information either in outline or pyramid (so-called hierarchical) form.

The Brainstorming Process

Whether you're a student who is preparing to complete a written assignment in mathematics, a professional engineer struggling in writing who needs to write a business report, or an environmental science professor attempting to adhere to a journal's submission requirements by putting a physical experiment into words as a completed research article, brainstorming can be a daunting endeavor. Typically, the individual stares at a blank page, not knowing what to write or what to do next. While an intimidating situation is not a good thing, we, as beginning writers, need to change our mindsets about the initial brainstorming period; rather than thinking of it as a stressful tedium, consider this initial experience in a positive light. The aspiration for people who are struggling is to find ways to become liberated from the difficult circumstances that contribute to their labors. Struggling situations could be when someone simply makes a mistake, puts ideas into words after a period of brainstorming, or is limited for a variety of reasons in finding an answer to a question or reaching a goal and is therefore required to discover ways to overcome these limitations. So, think of this initial brainstorming phase as a positive, motivating experience in which producing errors, writing nonsensical phrases, expressing irrelevant ideas, or making a grammatical slip is simply one part of the process toward reaching a goal. In fact, brain research has shown that making mistakes, whether consciously aware that an error was produced or not, contributes to neural growth and an overall increase in cognitive activity (Moser, Schroder, Heeter, Moran, & Lee, 2011). When I explain this to my students, many of whom are pre-service and in-service teachers, the comments I receive from them are similar to those that Boaler (2016) received, namely, most of them think that brain growth might exist when the individual is aware of the mistake produced. But, as Moser and his colleagues demonstrate, neurological growth also occurs when the individual is not aware that a mistake was produced. So, the objective when beginning in the writing process is to identify strategies of brainstorming that will successfully allow the individual to put ideas into words, words into sentences, sentences into paragraphs, and paragraphs to an entire manuscript.

So, then, what are some of the strategies that emerging STEAM writers use when brainstorming? What follows is a list and description of four strategies for engaging in the brainstorming process.

Strategy 1: Empty Your Thoughts on Paper

This strategy sounds strange, but it works for many burgeoning writers. The first thing to consider when using this strategy is to make a list of all the ideas that you can think of about your subject. It is not important if you write in complete sentences at this point; simply words and phrases will do, and it is all right if grammar and spelling is not entirely correct. The point of doing this initial exercise is to make as many connections that come to mind as possible and not to be too concerned about your written language seeming weird, unrelated, or even unscientific in tone at this point. Too much concern about delivery or tone runs the risk of omitting some crucial ideas that would otherwise be included in the text. Spend at least 20 minutes on this task. If you feel momentum or continuous flow of ideas, work for 30 minutes or for any amount of time until you believe you have enough written content to begin organizing your ideas.

Strategy 2: Make a Rough Outline

While I have a particular affinity to Strategy 1, mainly because it generally works for me, for other writers, it might simply be easier to start by making a preliminarily rough outline first. In other words, outlining, rather than freestyle writing (Strategy 1) for many writers is the way they engage in brainstorming. As soon as an idea comes to mind, the writer using this strategy will start organizing or creating some form of semblance of their ideas from the start. The main idea, then, is to begin by organizing content as you think of it.

Strategy 3: Create a Graphic Organizer

Strategy 3, creating a graphic organizer, I think, is a popular way of brainstorming that essentially links Strategies 1 and 2. Very new pre-service teacher education students are often intimidated by the term "graphic organizer"; something about it sounds scary or overly technical or scientific, involving byzantine or overly complex details. But this is not the case at all. I find that most things that students find intimidating are actually things that make us understand content more easily—things that create order and clarity. Such is the case of the graphic organizer.

The graphic organizer goes by many titles; *concept map, cognitive map, cognitive organizer, knowledge map, advance organizer, concept diagram,* and *word web* are a few of the more common designations. In general, a graphic organizer is a visual representation that allows the learner to recognize the relationships among concepts and how many of these concepts often interact. This instructional graphic organizing tool comes in a variety of formats, which include, but are not limited to, loose webs that contain central hubs, tables, or structured grids. The intention of use is to help students, STEAM writing students in our case, process information they have gathered and organize their ideas in a more coherent way. According to Bromley et al. (1995), graphic organizers generally are designed to follow one of four patterns of knowledge: hierarchical, conceptual, sequential, and cyclical graphic organizers.

Hierarchical graphic organizers use principles of rank of importance to help students break down a concept. These organizers have been successful study aids and scaffolds for writing projects. One can think of the Triple-Decker Essay Planner as an example of a hierarchical graphic organizer. This planner serves as a way of helping STEAM writers produce mini-essays as paragraphs—especially if they are trying to convince the reader of their position. The Triple-Decker Essay is in the following form: Introduction; Main Idea 1; Supporting Details 1; Main Idea 2; Supporting Details 2; Main Idea 3; Supporting Details 3; Conclusion. In organizing their ideas—from a central statement to convincing arguments (at least three), each of which include supporting details, this format allows the writer to move systematically from one level of information to the next.

Like hierarchical form, conceptual organizers, such as word or phrase webs, provide a format that itemizes the key elements of the main idea. As an open-ended format, conceptual organizers allow writers to generate ideas for a topic during the early stages of research—before the writing itself takes place. Conceptual organizers in the form of concept maps help the writer think hierarchically, not in terms of superiority, but in terms of priority of importance to the central thesis. For example, in Figure 3.1, one of my students created a cognitive map in the form of a word web that focused on the topic of "living things" for her term paper. She started off by producing key components to living things that were fresh in her mind: Classification, Life Functions, Biochemistry, and Cell Theory. She then studied each of these attributes in more detail and then constructed branches that pertain to each. For example, for Biochemistry, she began to differentiate between inorganic and organic compounds. She then found that organic compounds can then be subdivided much further. And so the cognitive map continues in this branching out format. While used more often in the social sciences, the "Cause-Effect Blossom" can also be applied to research in STEAM. The blossom's central space allows the writer to record what is believed to be the cause or stimulus and the leaves of the blossom can be hypothesized effects.

FROM BRAINSTORMING TO WRITING CRITICALLY IN STEAM 35

Figure 3.1.1: Two Levels

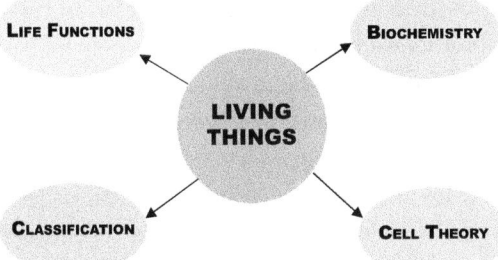

Figure 3.1.2: Three Levels

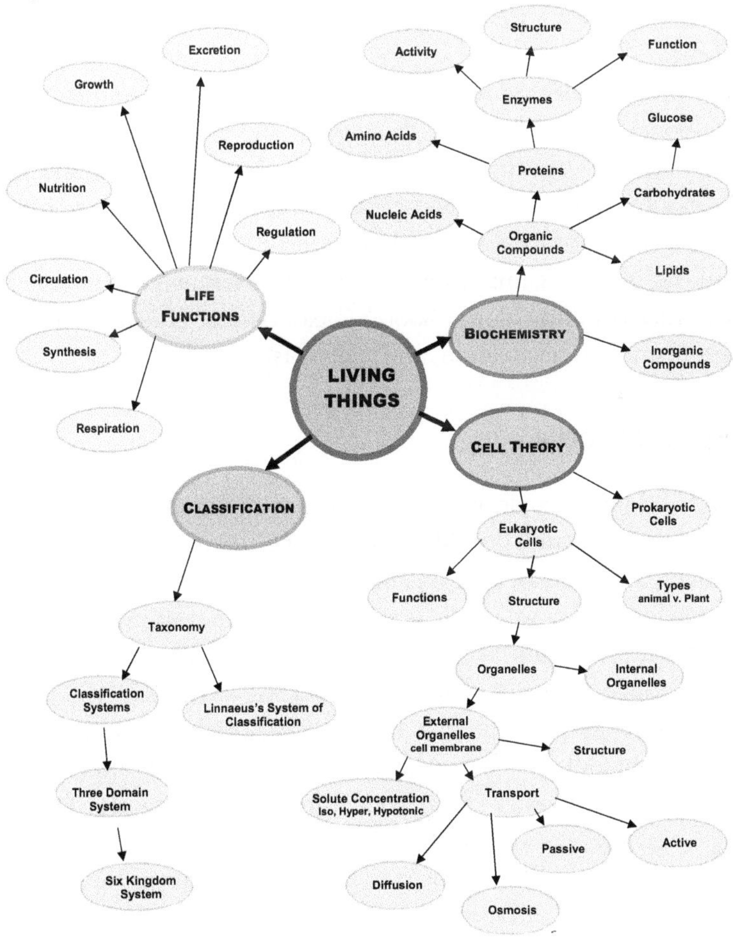

Figure 3.1. A Student-Produced Cognitive Map on the Topic of "Living Things"

Perhaps more relevant to STEAM brainstorming, might be sequential organizers, which are often used to indicate time-order relationships. Given time-order, this organizer is best viewed linearly as they can help STEAM writers link causes and effects and identify problem-solution relationships. Points or events on this time-order line can be as few as two to as many as needed. In this organizer, writers identify key occurrences, experiences, or events, in chronological order, which leads to a specific result. Cyclical organizers are used to show an ordered series of events that are part of a repeating pattern. The circular structure of these graphic organizers helps writers present, in order, each element, cycle, or succession. As writers complete the organizer, they discover that each time the sequence of occurrences completes itself, it begins again.

Graphic organizers are rewarding in that writers can arrange their ideas, possibly unorganized at first, but then more and more systematized as time progresses. Visually appealing and accessible to both struggling and advanced students, graphic organizers enable writers, STEAM students and professionals to connect prior knowledge to newly identified concepts, integrate language and thinking in an organized format, organize writing, and engage in mid- to higher-order thinking along Bloom's Taxonomy of Educational Objectives (application, analysis, evaluation, and synthesis).

Strategy 4: Moodle

Yes, moodle! Many people might have heard of "moodle" as the acronym MOODLE, which stands for Modular Object-Oriented Dynamic Learning Environment, software for programming. But that's not what we mean here. Moodling is something we all do, and we might not even know it when we're doing it. In fact, we might even be moodling before, after, or in between any one of the above strategies. Moodling happens when our minds begin to wander in directions that you may not have originally intended. Albert Einstein probably moodled as he sat in the streetcar in Bern and watched the Zurich city clock, an event that made him contemplate the ideas behind relativity—soon to become the theory of relativity. One of the first instances of the meaning of moodling in the context of the writing process was in 1938, when Brenda Ueland, in her book *If You Want to Write*, said "The imagination needs moodling,—long, inefficient happy idling, dawdling and puttering" (pp. xx). On his website, Eric Grunwald equates moodling with a sort of daydreaming session with a pencil or pen in one's hand whereby the writer puts imaginations on paper.

The fascinating thing about moodling is the idea that, perhaps more than each of the three previous strategies, it can help writers produce some of the most interesting, innovative ideas. I've outlined six items that can help writers, especially those who may be a bit more experienced than those just starting off, engage in moodling as they brainstorm their ideas.

1. Don't limit moodling time. If it lasts a half hour, an hour, or longer so be it. This is a time of extraordinary contemplation that may turn imagination into reality or science fiction into science.
2. Sit in a comfortable chair in a room that is conducive to study and introspection. While pen or pencil and paper is the ideal setting, a computer keyboard will do.
3. Begin to daydream and reflect upon your initial thoughts. Write down the most intriguing thoughts that you believe might be essential for your text. But don't force any thoughts to come out. If you feel as if your mind is producing thoughts, but do not feel these thoughts should be written down, then don't write them. Only start writing when you feel inclined to do so. Sometimes, when moodling, you might feel as if you're off topic, but that is all right. But don't start digressing to another activity. Focus on the present writing activity for as long as you are able to.
4. As indicated earlier, precise grammar is not of utmost importance during brainstorming. Unless you've produced a great sentence that you want to hone, avoid painstakingly checking for surface area errors, like grammar or mechanics, at this point. That time will come when you're in the process of writing and certainly by the time the process of editing and revision comes around.
5. Parallel to the first item, it will be important not to stress over the fact that you might not have written anything for a lengthy period of time— 20 minutes, 40 minutes, or even an hour. This is still a time of imagination and contemplation, when one is engaged in the deepest thinking. In other words, not writing does not mean not working. As Brenda Ueland has alluded to in the earlier quote, what you are contemplating about today is the creative part of the writing process; the organizational, grammatical, and syntactical structure that follows when one puts pen to paper or fingers to keyboard is simply the written record of

one's written creativity. At the same time, she emphasizes that if one is not writing anything due to fear or feeling "stuck," then pushing oneself to write would be the best remedy.
6. Alluding to the Einstein example above, when you develop the moodling habit, it might be a good idea to allocate a half hour to an hour each day for moodling. Doing so will likely lead to some really interesting ideas on which to write.

From Ideas to Words, Sentences, and Paragraphs: The Outlining and Planning Stage

At this point, the paper, or screen, is filled with a number of words, phrases, or sentences—or even pictures or diagrams. So, while brainstorming has begun to cease, you are now set out to start thinking of what to write. But before doing so, it will be important to take your words, phrases, or sentences, and find a way to organize them into an outline, one that is more formalized than the brainstorming outline mentioned above. This is a pivotal step because it will be difficult to jump into writing automatically given that your ideas are in their prototype and not yet developed. The formal, detailed outline will help develop your ideas into more coherent text. So, struggling a bit to develop a coherent outline at this point is actually a good thing. While not the most attractive part of the writing process, it is possible to increase motivation in developing an outline. One way to become motivated in this seemingly toilsome process is to consider this task as one that requires just as much thinking as the brainstorming itself. After all, your ideas are not yet sequenced, organized, or put into some sort of semblance. Engaging in this activity requires, still, a good deal of contemplation and forethought. Moreover, because it serves as a strong foundation for any writing project, the outlining process will make the initial draft much easier to write.

There are many tools at our disposal that we can use to do this. The first tool we will examine is ordering ideas from most general to the most specific—that is, in descending order of generality. We can use this tool both as outlining themes (e.g., in a literature review) or the outlining of items or categories, as in a Method section that includes participants, design, procedure, and the like. Here is an example of the first type in which I have written an outline for a literature review on research based on the theme of constructive free play:

I. Play Matters: Constructive Free Play (literature review)
   A. Ancient to Modern Conceptions of Constructive Play
      1. Aristotle's view on play
      2. Friedrich Schiller and play as a conduit to liberty
      3. Karl Groos (1901) and comparison of play in primates
      4. Johan Huizinga (1938/1955) and play for its own sake
      5. Jean Piaget (1951); Play for intellectual development
      6. Lev Vygotsky (1933/1966); Play as social cognition
   B. Contemporary Interpretations
      1. B. Sutton-Smith (2009)
         i. Ontological and epistemological underpinnings of play
         ii. Play and its origins in STEM
         iii. Self-testing play
      2. T. Henricks (2020)
         i. Play as a pathway to experience
         ii. *Communitas*
         iii. Psyche, body, environment, society, and culture
      3. Hirsh-Pasek and Golinkoff (2008)
         i. Cross-cultural conceptions of play
         ii. Play and its centrality to cognition
         iii. Play as the work of children
   C. Playworlds
      1. Daniil Elkonin (2005)
         i. Psychology of play
         ii. Play as instinct vs. higher-order cognition
         iii. Anti-dualism
         iv. Play and realism for the child
      2. Gunilla Lindqvist (1995)
         i. Development of PlayWorlds
         ii. Aesthetics of play
         iii. Play and higher-order thinking
         iv. Cognition, emotion, imagination, and creativity
      3. M. Fleer (2020)
         i. STEM begins in infancy
         ii. Engineering PlayWorld model for children
         iii. Motives and motivation for engineering play
         iv. A model of teaching science in play-based settings

In reviewing this outline, notice that it has been assembled in a more or less orderly way that was constructed with forethought. In other words, this completed outline is based on a more embryonic outline in which I organized the ideas described after each of the roman numerals. So, in this case, everything that follows "I. Play Matters: Constructive Free Play" is based on this very content. The next order of business is the list that follows each of the upper-case letters, "A, B, and C,"—namely, "Ancient to Modern Conceptions of Constructive Free Play," "Contemporary Interpretations," and "Playworlds." The main benefit of the so-called Outline Strategy is that it allows the writer to assemble their ideas in a more or less coherent way from the start so that it is a more expedient, and possibly easier, way to begin the writing process. A drawback, though, is the possibility that the writer might end up missing the opportunity to think of all the important ideas that could be included, at least in the initial draft of a budding and promising STEAM paper.

There are several benefits to the STEAM writer when outlining. First, the finished outline shows that the writer's ideas are now arranged in a more logically sound way than they were prior to the formation of the outline. The outline is the foundation of what's to come. Just as it is important to practice scales, arpeggios, and, at times monotonous etudes when practicing a musical instrument, so too is it important to develop an outline when writing a paper. Like that of the musician, the foundation for the writer reduces anxiety to the point that there are fewer and fewer things to keep in mind when one does actually get to the writing part. Second, brainstorming doesn't simply end with the conclusion of the brainstorming process that occurs prior to outlining; in fact, during the outline, the writer oftentimes discovers links that might have been overlooked when writing down ideas during brainstorming. It's almost as if one's mind is editing and adding new insight to what has been already jotted down or typed out. Third, outlining is a timesaving process that can reduce the number of future instances of reorganizing, rewriting, and reediting. And fourth, with a completed outline, the author begins to write at a more advanced and progressive level of thinking that helps build upon the foundation.

Drafting

So now we're at the stage when we begin to take our organized outline and start putting words together into sentences, sentences into paragraphs, and ensuring that each member of the sequence of paragraphs corresponds to the preceding paragraph and provides a good segue into the next. But the job of writing during the drafting stage is one that involves a bit of multitasking because the writer needs to consider how to get novel and, at times, intricate and complex ideas on paper, and, at the same time, to write them in the most optimal way with elegant style, which includes good grammar and mechanics. Clearly, writing can seem daunting at this point because these two tasks comprise two different cognitive processes.

On his website, Grunwald (n.d.) suggests the solution of separating these tasks into distinct steps. He argues that the first draft should simply focus on ensuring that one's ideas are roughly set into sentences, and not having the preoccupation of grammar, spelling, or specific vocabulary. Focusing on initially putting ideas into sentences is important because the ideas at this point are already outlined, which means that the writer is not working from scratch and therefore is building on something with which one is already familiar. And even more important, this is the drafting stage, indicating that after the outline is the first draft; in other words, the first go-around is the first draft of subsequent drafts—it's not the final product; you will revise and edit it later. It is not the case that you would save time on the first round by constructing perfect sentences in order to eliminate the steps of revising and editing. Thinking this way might actually increase stress and increase the amount of time that it will take to produce a quality paper. The key, then, is to focus on putting your ideas into words and sentences. It might be rough at first, but it will improve with the next two steps. The following list of steps should help the STEAM writer during the drafting stage.

1. Be sure to have a copy of the outline that was produced in the previous stage as you set out to complete the first draft. The outline should either be written or printed out on the desk or on the right or left side of the computer screen with the initial draft on the other side.
2. While reviewing the outline, simply start writing and keep writing. Again, the fine details of grammar or mechanics will be dealt with later.
3. When drafting, write in complete sentences—that is, do not write in fragments or string clauses together to form run-on sentences. Try to use the best grammar possible without focusing too much attention on it. Use the best wording as possible and transition words when linking sentences and paragraphs.
4. Don't ponder on a particular sentence or word at this point. Pause for a few seconds, but then move on.

When you've reached the point that everything seems to be down on paper and in complete sentences and paragraphs, then we will consider the revising stage.

Revising

At this point, one might not see the distinction between revising and editing. But it is important to identify what makes them different. Even more important is to revise first and edit later. But what does it actually mean to revise? And how does revising differ from editing?

If we break down the word "revise" into its Latin etymology, we get the prefix re-, which means "another time" or "again." The root of the word, "vise," comes from the Latin, *videre*, which means "to see." So, "revise" means literally to see again, or "re-see." In this regard (no pun intended for those of you who are French readers), we are seeing our draft again, and possibly a third or fourth time, or however many times are needed. When we revise, we are paying greater attention to the global aspects of the paper, namely, the organization and structure, and even more important, ensuring that the content is clear for readers to understand. The more surfacy aspects of writing, like spelling, grammar, syntax, and usage, are held off until the next and final stage. From a logical standpoint, it will not make sense to edit a draft before revising it because, during the revising stage, whole words, phrases, sentences, and even paragraphs might be changed, moved to another location, or outright eliminated from the paper. Therefore, editing prior to revising doesn't make very much sense.

If you recall from the previous section on developing an outline, newer ideas spring forth from the original ones that were penned while brainstorming. Similarly, revising allows the writer to reconsider certain ideas to see, first and foremost, if they make sense, and secondly, if anything that was written sparks a new set of ideas that can again recenter one's claims, restructure arguments, or bridge gaps in a more coherent and understandable fashion. Finally, it is essential to have someone else read the paper. Doing so provides a fresh outlook in terms of what works and doesn't work in the paper. I find that this is one of the most helpful aspects of the revising process because it allows the writer to recognize and appreciate someone else's position on the topic about which is written.

Editing

At this point, we arrive at the end point of the pen-to-paper or fingers-to-keyboard part of the writing process—in other words, in terms of generating the content and organizing the paper. Now it's time to edit the new work. Again, note that editing occurs as the last stage because during the revising stage, whole sentences and even paragraphs will be changed or omitted altogether. The logic is that it doesn't make sense to edit words or phrases, or even sentences or paragraphs that might be eventually eliminated, modified, or moved to another place. So, this way, the writer will only need to edit once.

Prior to editing, print out a copy of the saved paper that you just wrote. I'm not sure about everyone's feelings about which format to use when reading a copy of a newly written paper, but I personally like to print mine out, double-sided, when I begin to edit. That's just me. But I completely get it when someone tells me that reading the paper on the screen will suffice. Doing so will certainly help the environment. The printout, however, allows me to write with pen all over the paper. Screen editing, on the other hand, requires a great deal more time, not less, to edit; when editing, it's not the typing that takes the time. Rather it's the identification and the highlighting of words and phrases that need to be modified or added to. Further, it's easier for most individuals to identify errors on actual printed paper than it is for them to do so on the computer screen.

When editing, check for spelling, incorrect article usage, word order, and plural and singular nouns. For quick assistance, consult Appendices A through C to help you with issues pertaining to commonly confused words (Appendix A); most commonly misspelled words (Appendix B); parts of speech, an example of a diagrammed sentence, verb tenses, and rules for subject-verb agreement (Appendix C). Read through the paper at least once for each of these types of errors checking specifically for that problem; that is, read through for spelling, then read through for subject-verb agreement, and so forth. Doing so will increase your experience in improving grammar and usage, and also decrease the number of these kinds of errors during drafting. If there is any uncertainty during the editing process, consult Appendix D to determine which manual of style you will be using, and be sure to acquire this manual as it will contain all the necessary rules and protocols for ensuring that your paper is edited and formatted correctly. Assuming the paper is saved somewhere—on a hard drive, thumb drive, cloud—be sure to make the edits and save the newly edited manuscript.

Conclusion

In summary, we went from brainstorming to the final written product within the scope of this chapter. But of equal importance is the fact that each STEAM discipline has its own set of criteria that govern how a manuscript is to be organized and formatted. Moreover, each STEAM discipline requires its students, experts, and professionals to write different kinds of written documents for a different set of purposes. An engineer, for example, will be required to write business case documents while the biologist, chemist, geologist, or physicist will need to write up research papers with an entirely different nomenclature and type of organization. So, too, is the case for the writer of mathematics, who will need to write other kinds of papers. Thus, the next part of this book is all about specific writing, editing, and publishing conventions that are unique to each of the STEAM disciplines. We will discuss the STEAM disciplines in the order of S, T, E and A, and M—science, technology, engineering and architecture, and mathematics.

· 4 ·

WRITING CRITICALLY IN THE NATURAL SCIENCES

Stephen Farenga, Professor of Science Education and Program Director of Undergraduate and Graduate Secondary Science Education at the City University of New York, has maintained that the primary job of the science educator is not necessarily to produce young scientists; rather, it's to teach our students to think scientifically, that is, to be scientifically literate (Farenga, 2000). Indeed, this is not to say that professional scientists will not be among those who are learning scientific literacy in our classrooms. They already are! But the vast majority of our students will have interests in non-science-related professions. For example, a scientifically literate student is one who knows how to interpret data in order to determine the accuracy of a politician's statement or a news report. Thus, it is of utmost importance that our students, from preschool to Grade 12, have optimal science-related experiences because whatever profession they end up in, it will undoubtedly require scientific thinking.

If you are a student of science or a science teacher, you will undoubtedly engage in some form of writing that involves knowledge and expression of science content. Indeed, students will eventually need to submit science laboratory reports and research papers. Moreover, as they accumulate data, science researchers, whether student or professional, will need to know the general skills of writing in science when they jot notes down in notebooks and journals.

Before we delve into the topic of writing in the natural sciences as well as the general form of STEM research papers, we will need to differentiate between scientific writing and science writing.

Scientific vs. Science Writing

It is common to see and hear the terms "scientific writing" and "science writing" interchangeably. After all, the words "scientific" and "science" are almost the same, so they seem as if they have the same denotation. However, the meaning of these terms is different and each has its own distinct definition.

Scientific Writing

Scientific writing includes papers, books, periodicals, and other manuscripts that are based on science content in which the author uses specialized, technical scientific language and terminologies (Alley, 2018). From a cognitive perspective, Vygotsky (1986[1934]) referred to scientific writing as a writing genre based on scientific concepts, also called conventionally systematic concepts. Moreover, one will notice that scientific writing tends to be pithy. In other words, the community of natural scientists discourages elaborate, florid writing. Rather, they encourage concision and economy of expression. In this regard, it is important to avoid certain common phrases that are not only elaborate, but also redundant. For example, instead of "has the ability to," write "can" instead. In avoiding redundancy, write "note that" instead of "it is important to note that." The following list contains more recommendations for phrases to be avoided along with their improved counterparts (See Table 4.1).

Note that economy of expression can involve whole sentences as well. For example, "Antibodies were added to each individual sample for labeling" should read "Samples were labeled with antibodies" instead. As a second example, "The resulting liquid was blue in color" should be "The liquid was blue." And last, "Various modifications of the procedure have recently been developed" is too wordy. A more concise version is "The procedure was recently modified."

Scientific writing requires the writer to use specific nomenclature and terminology correctly. To this end, the scientific writer will need to become acquainted with commonly used scientific taxonomies. The Swedish botanist Carolus Linnaeus solved the problem of naming organisms and species, especially when different scientists from different parts of the world attempted to talk about organisms that assumed different names originating in very different languages. In 1758, Linnaeus proposed a system for classifying organisms in his treatise entitled *Systema Naturae*. In this system, each species is assigned a two-part name, also called binomial nomenclature, and, for the purpose of universality, names are all in Latin. The first part of the scientific name is the genus, and the first letter is always upper-case. The second part is the species name, and when put together, the entire name is italicized. For example, *Homo sapiens* refers to humans. Upon the publication of Charles Darwin's book, *On the Origin of Species*, evolutionary history of organisms became an important part of classifying organisms. Today, sophisticated techniques such as DNA sequencing are essential ways in which scientists engage in classification. So, each genus contains species that share common ancestry. For example, because wolves (*Canis lupus*) and coyotes (*Canis latrans*) arose from a recent common ancestor, they are placed in the same genus, namely, *Canis*. The following are common scientific terminologies:

- Species and Latin derivatives: written in italics with an upper-case first letter of the first word (e.g., *Danio rerio*)
- Human genes: written using all upper-case letters and in italics (write peroxiredoxin family member 1 as *PRDX1*)
- Mouse genes: upper-case first letter only (e.g., *Sta*)
- Human proteins: upper-case and non-italicized letters (e.g., ADH3)
- Mouse proteins: same as mouse genes but no italics (e.g., Sta)
- Restriction enzymes: combination of italic and non-italic letters (e.g., *Eco*RI).

The following example of scientific writing demonstrates the use of highly technical language only to be found in research literature of biochemistry (Field, Graf, & Link, 1952):

> ... Methylxanthines, caffeine, (1, 3, 7-trimethylxanthine), theobromine, (3, 7-dimethylxanthine), and theophylline (1, 3-dimethylxanthine) when given orally, induce a state of hyperprothrombinemia and reduce the extent of the hypoprothrombinemia induced in the dog and rat.

In the next section (Science Writing), notice a passage on the same topic but written in more explicable form.

Science Writing

In contrast to scientific writing, in science writing, the author is not compelled to use technical language. Rather, science writers take specialized written texts and adapt the science concepts therein in a way that introduces non-specialists and learners to new science concepts and procedures. Science writing passages are composed in non-technical language that allows for greater accessibility to the wider public. From a developmental standpoint, Vygotsky (1986[1934]) would consider science writing as a writing genre based on everyday, spontaneous concepts. Unlike passages in scientific writing, those in science writing tend to be more explicable in order to better equip laypersons and students with the tools needed to learn about science. Therefore, science writing is more detailed and explanatory than scientific writing is. Also, while economy of expression is a good rule to live by in science writing, writers tend to include more explanation using layperson terminology.

The following example from Farenga, Ness, and Hutchinson (2008) comes from a passage of science writing. When reading it, unlike its analogous passage in the previous section (Field, Graf, & Link, 1952), notice its slightly more apparent fluidity and greater use of everyday, detailed information in explaining science concepts for the layperson—concepts that would otherwise be written in terse, specialized language for biologists, biochemists, and researchers of veterinary medicine:

> Chocolate is dangerous for dogs. It contains stimulants called methylxanthines, which have a similar effect to caffeine. Chocolate may cause excitability, vomiting, excessive panting, thirst, heart arrhythmia, seizures, and can be fatal when ingested. Like chocolate, onions are toxic for your pets. Onions cause irreversible anemia, which is a blood disorder that damages the red blood cells in dogs and cats. The ingestion of onions may cause an even greater threat to cats because their red blood cells have shorter life spans. Animals may demonstrate decreased heart rates.

Comparisons of Scientific and Science Writing Passages

The basic difference, then, between scientific writing and science writing is that while scientific writing requires the use of technical, specialized language, oftentimes including jargon (that will need to be defined) as a means of scientific expression intended for audiences of like-minded and experienced professional scientists in formal settings, science writing is based on non-technical, jargon-free language intended for the student or science layperson.

As an important aside, one might question the need for scientific writing if few people are able to understand or decode its language. However, challenging the need for scientific writing is baseless and without merit. Note that without authors—professional scientists—who engage in scientific writing and communication, science writing would not have much, if anything, on which to base its conclusions. One of our goals as educators is to connect everyday writing with conventionally systematic writing (Sawyer & Liggett, 2012). A clear example is Albert Einstein's (1905) original publication of the theory of special relativity. Einstein's objective in his article starts out this way:

> We shall raise this conjecture (whose content will be called "the principle of relativity" hereafter) to the status of a postulate and shall introduce, in addition, the postulate, only seemingly incompatible with the former one, that in empty space light is always propagated with a definite velocity V which is independent of the state of motion of the emitting body. These two postulates suffice for arriving at a simple and consistent electrodynamics of moving bodies on the basis of Maxwell's theory for bodies at rest. The introduction of a "light ether" will prove superfluous, inasmuch as in accordance with the concept to be developed here, no "space at absolute rest" endowed with special properties will be introduced, nor will a velocity vector be assigned to a point of empty space at which electromagnetic processes are taking place. Like every other electrodynamics, the theory to be developed is based on the kinematics of the rigid body, since assertions of each and any theory concern the relations between rigid bodies (coordinate systems), clocks, and electromagnetic processes. Insufficient regard for this circumstance is at the root of the difficulties with which the electrodynamics of moving bodies must presently grapple (p. 891).

This excerpt is an example of scientific writing. Now, examine the following excerpt on the same topic of special relativity:

> Maxwell's equations claim that the speed of light is a universal constant, 186,000 miles per second, for every observer in the universe. But this statement leads to a stunning paradox. In our everyday experience, velocities are additive. Different observers describe the same events differently in different frames of reference. Albert Einstein thought about this situation and realized that our common sense about the way velocities add could be wrong, since our everyday experience does not deal with objects traveling anywhere near 186,000 miles per second. Einstein began to think about relativity while looking out a streetcar window moving away from a clock tower, just as the clock was about to strike noon. He imagined what he would see if the streetcar car increased its speed, faster and faster, close to light speed. He realized that he would be "surfing" on the light waves that carried the information

that it was noon. The timepiece he carried with him would tick away seconds. The clock tower, however, would appear to slow down and stop. The astonishing conclusion is that the measurement of time, like the measurement of motion, is relative to the onlooker's frame of reference.

In contrast to the first excerpt, which demonstrates scientific writing, the second exemplifies science writing. Given that they share the same basic information about special relativity, notice some of the parallels between the two excerpts.

Now that we have distinguished between scientific writing and science writing, we will examine a set of two essays in science—one organized as scientific writing and the other as science writing—based on the same topic. Determine which one of the following two short excerpts is an example of scientific writing and which one is an example of science writing.

Excerpt 1

Cyanobacteria are highly promising microorganisms for biological photohydrogen production. The review highlights the advancement in the biology of cyanobacterial hydrogen production in recent years. It discusses the enzymes involved in hydrogen production, viz. hydrogenases and nitrogenases, various strategies developed by cyanobacteria to limit nitrogenase inactivation by atmospheric and photosynthetic O_2, different biochemical and physicochemical parameters influencing the commercial cyanobacterial hydrogen production and the methods opted by different researchers for eliminating them to obtain maximum and sustained hydrogen production (Madamwar, Garg, & Shah, 2000, p. 757).

One might find this excerpt in journals of the likes of *Nature*, *Science*, or *Plos One*, which are peer-reviewed journals with very competitive manuscript acceptance rates.

Excerpt 2

Cyanobacteria are oxygenic, photosynthetic, micro-organisms and are widely distributed over a diverse range of habitats. Remaining in the oblivion, uncared and unrecognized, it has shot into fame and popularity owing to a host of their innate properties that make them ideal organisms for use in a variety of ways to meet our needs and to promise us a bright future. However, it is essential to explore new species of cyanobacteria existing in the nature, isolate and purify it and subsequently establish a collection which then could be a door for the biotechnological exploitation (Shah, Garg, & Madamwar, 2000, p. 175).

One who thinks that this excerpt is one based on science writing is correct. That is, this is the type of excerpt that one might find in a science journal for practitioners, such as the journals *The Science Teacher* and the *Journal of College Science Teaching*. These journals are also peer-reviewed publications; however, they are geared for science teacher practitioners who need to take seemingly complex ideas and structures and present them to students.

Reading and Reviewing Science Content

Let's get back to the idea of scientific literacy and thinking scientifically. Our goal, then, is to ensure that all students, regardless of background, are able to think, read, and write as scientifically informed citizens. While it is true that some of these students will enter fields in science, those who do not can gain a great deal of scientific knowledge.

Research Papers in Natural and Social Science-Related Disciplines

The science research paper: it's a form of communication to which the overwhelming majority of STEM specialists aspire. But it's not only a form of communication; the research paper is also a form of unearthing previously unknown information, much of which directly affects society as a whole. It is at this time, then, we turn to the research paper to identify its key elements that nearly all scientific investigators use when reporting on new findings and possibly turning their papers into articles in peer-reviewed journals. It turns out that the format of research papers in science does not really change that much when comparing natural science and social science research formats. So, what does the scientific research paper include that sets it apart from other papers or reports? It's important to keep in mind that the overall structure of a scientific research paper is somewhat formulaic in nature. In general, nearly all write-ups of research findings and the data used to support them have the following four main headings:

1. Introduction
2. Methods
3. Results
4. Discussion/Conclusion

When you have completed your four-sectioned paper, you will then extrapolate key points and collect these points for your abstract, which will be placed after your title and before your paper's introduction. In what follows, we will discuss what goes into pre-writing and, subsequently, each of these main headings in detail and provide a general overview of each of their important subheading topics so that you get the big picture as to how to go about structuring your research paper.

Pre-Writing Process

In the previous chapter, we devoted a significant amount of time focusing on brainstorming and how important it is to start putting ideas into words. We're going to use those recommendations as a launching pad for constructing the format of the scientific research paper. To begin, you know that you'll need to start writing your paper when you have a sufficient number of data to help you tell your story and be confident that you can support your findings with evidence. According to Schultz (2013), it can be helpful to think in terms of figures. The idea is this: If you lay out your figures in some semblance and order that makes sense to another person who examines your figures and can identify your main idea, methods, findings, and conclusions, then you seem to be on your way to begin writing up your research.

Before writing up your research paper, be sure to read the previous chapter on brainstorming. Consulting Chapter 3 and spending time organizing your ideas critically before putting pen to paper will hopefully ease any stress and save time later. That said, the following is a general protocol for the pre-writing process:

1. Review the research literature in your specific STEM area. What studies exist within the broader topic that you are investigating? What do they tell us that we can know? And how does your research help fill the gap(s) that is missing? Note that you will need to identify each reference for the citations that you include not only in the literature review but also throughout the paper as a whole.
2. Identify your audience. Will your audience include researchers in your STEM field? Researchers in other STEM fields? Students or laypersons? Also, what might these individuals' past experiences in the field be? This question is important because the answer to it will help you identify the appropriate terminologies for your readership at large.
3. Identify your research questions, the purpose of the study, and your hypotheses. Doing this will help you answer the question: "What

motivated you to pursue this research?" and "Why, then, did you decide to conduct it?"
4. Develop an outline. The information below should help you do this in a straightforward, concise manner. However, there might be some individual differences depending on the content of the research. For example, some Methods sections have subsections entitled "participants," "materials," "design," "instruments," "ethical considerations," and "procedure(s)." These subsections may not always be strictly sequenced in this way. Moreover, some studies will not have some of these subsections. Some papers, in fact, might combine materials and instruments into one section, and so forth.

It is common for many STEAM writers to have what is commonly called writer's block, a period of time when the writer is temporarily stuck or impeded in being unable to put ideas into words and formulate sentences and paragraphs. There are several things that one can do to rectify writer's block. First, be sure to have all your materials organized; this includes your notebook, figures, diagrams, tables, and references that are used to support your thesis. Second, pick up a copy of the journal to which you might be thinking of submitting a paper. Skim the articles to look for the structure, style, and general format of each article to get a sense of how you might go about structuring your own. Third, go to the website or initial pages of the paper journal and identify the author instructions. Each journal, no matter if your target journal is similar in subject matter to another journal, will have its own instructions for authors. This knowledge will help you formulate some of your initial ideas as you align your paper to the journal of choice. Fourth, when starting off, think small, not big. In other words, work on a single paragraph at a time. Then you will have a sense as to how your paragraphs will fit together in order of sequence. Fifth, don't worry if you're not sounding pedantic; you shouldn't be in the first place. Write down how you are thinking at any given time; don't write down what you think other people want you to write. Be your own author. Indeed, emulating great models is a good thing, but it's important for readers to listen to your own voice on a given topic and not someone else's. Sixth, if you've followed each of the preceding suggestions and you're still having difficulty in putting words on paper, then set deadlines for your work. This might help you stimulate ideas in a way that will unfold as you move on to subsequent paragraphs or sections of your paper. Seventh, save your work in several virtual locations. Do not save your important work solely in one location, such as a USB thumb drive. People often lose thumb drives and break them unintentionally. Save your work on both internal and external computer drives. One can also save work in a so-called cloud—that is, the practice of using remote network servers on the Internet to easily store, manage, access, and process data. Lastly, if you're working with a coauthor, share ideas with that person in a way that allows each of you to read the other's writing. If you're working alone, find a friend or colleague who might have the time to review your paper and provide constructive feedback.

Title

Clearly, the title of your research paper does not, and will not, tell the reader everything that needs to be told about your study. However, it should, indeed, summarize and encapsulate the general idea of what you're writing about. In this regard, it might be useful to consider four general things to keep in mind as you consider a good title for your paper. First, your title should draw readers' attention toward the overall theme of your article. The title should at least make readers' heads turn to at least the abstract, and possibly the whole thing. Second, and perhaps the most self-evident point, be sure that your title, in some way, states the main idea of the paper as a whole. One great way to ensure this is to identify key words that underscore or synopsize your main points. So, for example, in the scientific writing above, some key words might be "hypoprothrombinemia," "methylxanthines," and "caffeine." Key words in the science writing excerpt above might be "pets," "chocolate," "onions," and toxic [foods for animals]. Third, in terms of aesthetics, the title should be as compelling and persuasive as possible, yet, at the same time, clear and concise. And lastly, your title should make your research study unique in your field. There's no question that it is really fun to think about nifty titles. But the initial focus should clearly be on the substance of the write-up. The title should be one of the last, if not the last, item you address in the paper writing process.

Abstract

If the title isn't the last thing you deal with when writing your paper, then the abstract should definitely be. There's really no way that an author can appreciate the gist of the entire paper prior to creating the methodology, and most important, obtaining findings. The basic idea is that the abstract should include something about each of the main sections. The abstract, in essence, is a miniature rendering of the entire paper. The general parts of your abstract include the background of your research, a statement of your question or hypothesis, the research approach (e.g., experimental, empirical, etc.), results, and implications. These parts are not given equal weight in terms of length. I would emphasize here that the abstract is short; in my experience, I've never seen abstracts exceed 200 words. The typical range is anywhere between 75 and 200 words. If there is a plan to publish the paper, the specific journal to which the paper will be submitted will include a section for authors that must be checked for abstract word count requirements. Many STEAM writers believe that the abstract, due to its relative brevity, is the simplest part to write. It's not. The writer has to select the right words to exemplify the essence of the paper, which, on many occasions, can be very time-consuming. Table 4.2 discusses the general components of an abstract and its typical length.

Introduction

The introduction is an interesting section. It's the only section of the research paper that doesn't necessarily need a main heading. Assuming that it includes a literature review, oftentimes STEAM writers will use the term "Literature Review" as a heading. At other times, authors might begin with a so-called need of the study to show that the research question is very important to the field. The basic idea is to employ what is known as a funnel structure—going from the general to the specific. In other words, the literature review will start with general references in your field. You will then divide the general areas of the background knowledge into two, three, or four (or more) parts. For example, if the study is to determine the extent to which 4- and 5-year-old girls engage in everyday, spontaneous mathematical activities in the free-play center of the preschool, there are several intersecting areas that the author might wish to look at (Ness & Farenga, 2007). These areas include, but are not limited to, developmental psychology, mathematical cognition, early childhood education, play, and mathematics in the early years. The idea is to arrange the literature by examining each of these (or other) related areas from general to specific. It would seem as if "mathematics in the early years" would be that last part to examine before showing that your study is missing from the literature and your intention is to fill that gap. The key idea in this section is to avoid any research literature that falls outside the scope of the area that you are investigating. For instance, in the above example, I would avoid examining research on adolescent mathematics or aeronautical engineering because they are irrelevant to the topic at hand. The final few sentences (short paragraph possibly) of the introduction should include where the problem or missing research lies, and how the author is going to address this unknown. The author will then end the section by restating the research question.

Methods

Begin the Methods section by introducing the overall approach to the research. What research problem or question did you investigate, and what kind of data did you need to answer it? Is it quantitative, qualitative, theoretical, or historical? Depending on your discipline and approach, you might also begin with a discussion of the rationale and assumptions underpinning your methodology. In a quantitative experimental study, you might aim to produce generalizable knowledge about the causes of a phenomenon. Valid research requires a carefully designed study with a representative sample and controlled variables that can be replicated by other researchers. In a qualitative ethnographic case study, you might aim to produce contextual real-world knowledge about the behaviors, social structures, and shared beliefs of a specific group of people. As this methodology is less controlled and more interpretive, you will need to reflect on your position as researcher, taking into account how your participation and perception might have influenced the results.

Once you have introduced your overall methodological approach, you should give full details of the research methods you used. Outline the tools, procedures and materials selected to gather data, and the criteria that were used to select participants or sources. In other words, are surveys being employed, or is the research data going to come from an experiment or empirical study?

Next, you should indicate how you processed and analyzed the data. Avoid going into too much detail—you should not start presenting or discussing any of your results at this stage. In the methods section, you might include how you prepared the data before analyzing it (e.g., checking for missing data, removing outliers, transforming variables), which software you used to analyze the data (e.g., SPSS or R), or which statistical methods were used (e.g., correlation, two-tailed t-test, simple linear regression, analysis of variance (ANOVA), analysis of covariance (ANCOVA), or multiple analysis of covariance (MANCOVA)).

Your methodology should make the case for why you chose these particular methods, especially if you did not take the most standard approach to your topic. Discuss why other methods were not suitable for your objectives and show how this approach contributes to new knowledge or understanding.

You can acknowledge limitations or weaknesses in the approach you chose but justify why these were outweighed by the strengths. Lab-based experiments can't always accurately simulate real-life situations and behaviors, but they are effective for testing causal relationships between or among variables. Unstructured interviews usually produce results that cannot be generalized beyond the sample group, but they provide a more in-depth understanding of participants' perceptions, motivations and emotions. Remember that your aim is not just to describe your methods, but to show how and why you applied them and to demonstrate that your research was rigorously conducted.

Results

There is a plethora of terms we can use in place of "results." Indeed, the words "outcomes" or "findings" come to mind, too. That's exactly what the results are: findings or outcomes. This is the section in which the main findings are reported. If your research question requires more than one type of method (e.g., two experiments or a correlation and an experiment), report the findings in each of these as they appear in the Methods section. So, if the Methods discusses an experiment first and a correlation second, the Results should show the same sequence of findings. It is generally assumed that your Results section will have diagrams, figures, or tables. As indicated earlier, good pictorial depictions should tell the reader most of what is needed to be known (Coppens, 2016). But your text will explain the details; your text must point the reader to the data shown in the charts, figures, or tables. Most important, the Results is not the place to discuss conclusions, implications, assumptions, or comparisons with other research studies. These types of information are discussed in the next section.

Discussion/Conclusion

Now we're in the right place to talk about what our findings say about what you were set out to find in your research question(s). Unlike the introduction, where the direction was from general to specific, the Discussion should interpret your question—the specific—and begin to move toward the general, namely, how your question addresses your field at large. It is also in this section where you can finally compare your findings with those of other studies. Readers would like to know, for example, whether your findings corroborate those of others, or that your findings go above and beyond the extent to which other researchers were able to answer their specific questions. Discuss the implications of your research. That is, what impact does your findings have on a particular STEAM-related phenomenon? And how does this impact affect the community or society? Given these strengths, the Discussion is also the place to consider limitations, and possible unexpected findings. A conclusion—a paragraph or small section—should then end this section and the paper in general, one that restates your interpretation of the findings and possibly mentions possibilities for future research.

Summary

In this chapter, we have discussed key strands that stand out in importance for writing in the natural sciences. In short, the writer of science must be clear, succinct, precise, and logical. In addition, the writer of science must consider the audience for whom she is writing. The following are summaries of these key points:

> **Clarity:** Scientific research, whether in scientific or science writing, must be stated as clearly and simply as possible. Rambling, overly expressive, and embellished text is unsuitable for writing in science. That is, the findings of one's scientific narrative should be impossible to misinterpret.
> **Succinctness:** Scientific writing must be brief and pithy. In-depth concepts should not be made more complicated and confusing. We discussed the importance of avoiding redundancy. Moreover, avoid clichés as this language already possesses redundancy.
> **Precision:** When writing in science, choose words carefully. Another scientist or student of science should be able to repeat the experiment in a report or article. To be sure, the reader should know exactly how the results relate to each other and to the results of other scientific experiments or empirical studies. Avoid ambiguity and vague words and phrases by conveying the exact meaning of what was intended.
> **Logical Writing:** Scientific arguments in reports, papers, or articles must be logical. One's argument is the final claim, which is based on previous claims known as premises. So, in a sound argument, the premise must be true.
> **Audience:** Think about who will read your writing and why they will be reading it. One should ask: Will the reader understand my explanations? And even more important, do I understand my own explanations?

Writing Critically in Science: Some Brief Exercises

1. Shorten the prepositional phrases in the following sentences:
   A. Continuous treatment with steroids was necessary to prevent inflammation of allergies.
   B. The method of greatest efficiency included a number of steps in standard use.
2. Revise the nominalizations in the following sentences:
   A. Our tasks were the collection of data and the development of innovative products.
   B. Performance of the new protocol caused a 10% increase in the desired effect.
3. Identify and correct the redundancies in the following sentences:
   A. As a general rule, we like to keep our presentations brief and concise.
   B. They emptied out the flask before repeating the procedure again.
   C. In their study, Smith et al. (2000) focused specifically on three species that were similar in nature.
4. Replace the following roundabout phrases with one word:
   A. as a result of
   B. have an effect on
   C. at a slow rate
5. Rewrite the following paragraph to reduce wordiness:

In this review, an attempt has been made to provide a description of the nature of Alzheimer's disease (AD), including the causes, symptoms, and management of this condition. There is reason to believe that AD is an extremely common form of dementia that causes an increase in a number of difficulties with memory, an intensified escalation in issues with behavior or with conduct, and a gain in problems with thinking and pondering. In spite of the fact that the majority of patients show signs of the development of AD after they are older than 65 years of age notwithstanding, some patients receive a diagnosis of AD prior to that age. A quite progressive neurodegenerative disorder, AD is a disease with worsening effects over the course of a person's lifetime. The stages range in the amount of severity from an absolutely mild to severe stage until substantial interference with daily activities occurs. It was already reported earlier in the literature by several research investigators that in AD, amyloid plaque develops for the most part in the hippocampus, which is a brain structure that helps in the encoding of memories, as well as in other areas that are largely involved in our thinking and behavior. It is not known at the present time whether plaque causes AD or whether it is a consequence of the AD disease process. (226 words)

Table 4.1. Improving Redundant Phrases in Scientific Writing

Wordy Phrases	Same Idea Improved
A majority of	Most
A number of	Many, several, numerous
A small number of	Few
Absolutely certain	Certain
Actual fact	Fact
Add an additional	Add
Are known to be	Are
As a consequence of	Because
Ask a question	ask
at the present time	at present
At the same time	While
Close proximity	proximity
Completely filled	Filled
Consensus of opinion	consensus
Due to the fact that/in light of the fact that	Because
During the course of	During
Estimated at about/roughly	Estimated
Few in number	few
Fewer in number	Fewer
Filled to capacity	Filled/at capacity
For a period of	__ days, __ hours, __ years, etc.
For the purpose of	For, to
For the reason that	Because
Has the opportunity to	Can
In a routine manner	Routinely
In order to	To
In spite of the fact that	Although, despite
In the case of	For
In the course of	During
In the event that	If
In the near future	Soon
It is important to point out that	Note that
It is often the case that	Often

Table 4.1. Continued

Wordy Phrases	Same Idea Improved
It is possible that	May
It would appear that	apparently
Major breakthrough	Breakthrough
On the basis of	By
On the order of	About
Postpone until later	Postpone
Prior to	Before
Referred to as	Called
Repeat again	Repeat
Spell out in detail	Spell out
Subsequent to	After
To put it another way	In other words

Table 4.2. Components and Length of Abstract

Abstract Idea	Location	Length
Background	Introduction/Literature Review	1–2 sentences
Purpose/Hypothesis	Introduction/Literature Review	1 sentence
Research Approach & No. of Participants	Methods	1 sentence
Results or Findings	Results	1–3 sentences
Implications	Discussion/Conclusion	1 sentence

· 5 ·

WRITING CRITICALLY IN TECHNOLOGY

Before delving into discussion about writing in technology, I need to raise two initial points. First, note that in several instances (see Chapter 6), engineers will often refer to engineering and technology interchangeably. Today, the International Technology and Engineering Educators Association (ITEA, 2020) has transitioned from its past role into the present STEM context with a set of standards with a somewhat prolix title: *Standards for Technological and Engineering Literacy Defining the Role of Technology and Engineering in STEM Education (STEL)*. In the outset, *STEL* begins by defining "technology" as anything constituting the "built world"—hence the pivot role of engineering in STEM—and continuing into the present computer universe. In support, Kosky and colleagues (2015) defend this "engineer view" of defining technology as anything humankind that has been added to the world environment. They define engineering as the application of mathematics, the natural sciences, technology, and experience to create a system, component, or process that serves society. For example, *Engineering in Context*, by Sir Alan Muir Wood, the private engineer to the Queen of England, used the terms "engineering" and "technology" interchangeably in single sentences while referring to the same subject. Many engineers with whom I have communicated seem to indicate that they see the two terms as similar entities. Even the creation and development of computer programs through coding—what we might think of as computer software engineering—is a form of technology.

The second point that needs to be stressed is connected to the first; it has to do with the word "technology" itself. When writing in technology, we will not limit our discussion of technology writing solely to what many of us think of when we first hear the word "technology"—namely, devices with mostly silicon-based semiconductors embedded in them that allow for the relatively instantaneous production or completion of a specific task after a command is given—that is, what a personal computer, laptop, or smartphone does. For our purposes, technology ranges from the most archaic object to the most sophisticated one: from chalk and chalkboard to devices that write down your thoughts through speech recognition; from ancient sundial to nuclear clock, and the like.

That said, we will examine critical writing in technology by defining technology according to the definition propounded by the International Technology Education Association for the Technology for All Americans Project (ITEA, 2020): "Broadly speaking, technology is how people modify the natural world to suit their own purposes. From the Greek word *techne*, meaning art or artifice, but more generally it refers to the diverse collection of processes and knowledge that people use to extend human abilities to satisfy human needs and wants" (p. 2). In this chapter, then, we will discuss writing critically in technology and connections with engineering when necessary.

Developing Ideas in Technology

Developing new technologies is more than problem-solving; it also requires the individual to engage in problem posing or problem making. Moreover, this idea of problem posing leads to a problem that will need to be solved. And of course, to solve a problem, we engage in technology development to help us do that. So, what is problem posing exactly, and why would it be important to consider problem posing as a key component when writing in technology?

From a STEAM perspective, problem posing is generally defined to be a reformulation of a given problem or the ability of producing new problems or questions that need to be addressed. When STEAM writers pose their own problems, they can enhance their specific STEAM-based knowledge. Doing so stimulates higher-order thinking—i.e., analyzing, synthesizing, and evaluating—and improves computational skills from the perspective of the individual who can explore their own interests and motivations about specific

STEAM concepts through metacognitive activities. We can extend the notion of problem posing to problem posing education, which is a term coined by Freire (1970). For Freire, problem posing education is a critical method of teaching that underscores the importance of critical thinking for the purpose of liberating oppressed populations. Freire believed that a problem posing educational model is an empowering and inspiring alternative to traditional approaches of instruction that is, for the most part, based on the behaviorist paradigm that advocates classical and operant conditioning procedures. While many problem posing researchers would claim that Freire used problem posing as an alternative to the "banking model" of education, I would argue that he used problem posing as a direct indictment against the banking model, which only serves to subjugate and marginalize the knowledge of students. Problem posing, however, is extremely limited, if non-existent, in education, let alone STEAM teaching and learning. With few exceptions, the overwhelming majority of learners lack the vital skill to engage in inquiry that can be employed to enrich learning experiences. And, even rarer is the ability to engage in adaptive inquiry (Farenga, Joyce, & Ness, 2006). Rather than posing multilayered, manifold, and profound problems, most individuals pose very few problems, and those who do, pose narrow and superficial problems that only address the so-called who, what, when, and where problem types. The goal in technology writing and other written STEAM documents is to engage in deep questioning in problem posing—that is, the why, how, what if, and to what extent question starters. In sum, writing by problem posing represents the embodiment of how critical pedagogists address the essentials of critical writing (Steinberg & Kincheloe, 2018), particularly critical writing in STEAM (Farenga, et al., 2010). Problem posing theorists generally divide the development of one's problem posing skill into three phases: the initial instruction phase, problem posing phase, and addressal phase.

The initial instruction phase is one in which problem posing is semi-structured for the purpose of providing STEAM students with the opportunity to explore for inadvertent situations that need to be addressed as the teachers use scaffolding as a form of guided instruction. Mishra and Iyer (2015) refer to the initial instruction phase as "seed knowledge." Furthermore, initial instruction overtly contains hints or clues, which can encourage students to deal with questions that involve exploration and investigation. In the problem posing phase, the second phase, students pose questions based on the content they have investigated from seed knowledge. The purpose of the questions in the second phase is twofold: 1) to clarify any murky argument related to seed knowledge and 2) to further advance inquiry skills and discovery to increase understanding connected with seed knowledge. And third, in the addressal phase, the teacher attends to clarification questions initially and exploratory questions subsequently. Clarification questions are all the questions which require restatement of the subject matter that has explicitly been taught in any other previous lectures in the course. In contrast, exploratory questions are the questions that bring about the construction of new knowledge.

After problem posing comes problem-solving. George Pólya, an eminent mathematician and philosopher in the field of problem-solving, is credited for his four-step problem-solving method. Pólya, a Hungarian mathematician who immigrated to the United States before the Second World War and taught mathematics at Stanford University, was also interested in how mathematics was instructed to students prior to attending the university. Mathematics instruction was one of his interests during the late 1930s and early 1940s, at a time when he was developing algorithms, essentially computer code, for the United States military. During this time, Pólya was searching not only for algorithms that enable the functioning of various applications—he was also searching for the maximization and optimization of the application's efficiency. This interest led Pólya to examine methods of problem-solving that can be used not only for research, but for educational purposes as well. And so, Pólya's four-step method for solving problems evolved from his development of algorithms for, at the time, modern technologies. The profundity of Pólya's four-step method for problem-solving is based on its seemingly self-evident and intuitive structure:

1. Understand the problem that is given to you. The problem solver must be clear about what the problem entails and what type of solution is appropriate.
2. Devise a plan to solve the problem. In this step, it is assumed that the problem solver will need to tap and then identify prior knowledge as a means of formulating a strategy for solving the problem. It could entail the ability to know which algorithm or set of algorithms to use or perhaps to utilize one or more heuristics that will approximate a solution.[1]
3. Carry out the plan. This step simply calls for the problem solver to execute the plan that was devised in step 2. Again, the execution of the plan may involve anywhere from one instruction to multiple algorithms or heuristics in a chronological and orderly manner. It is in this step that the solution appears.

4. Check the result or solution. It is important to ensure that the solution that was found is one that is both valid and reliable. Students of all ages often fail to check their results. Doing so often runs the risk of erroneous solutions.

This approach is important in our context because it helps students identify ways to write that express clarity and consistency of understanding. The techniques involved in problem posing and problem solving are intrinsically connected with technological inquiry, not only in terms of conceptual development but in writing development as well.

Reviewing Technology Content to Putting Ideas into Words

It is necessary to stress at this point that the idea of writing critically in technology is not identical to the idea of incorporating technology into writing instruction. The point here is to provide examples of writing about technology in ways that demonstrate critical writing as well as connections to other STEAM disciplines. When writing about technology, you will need to arrive at a topic that is not too broad in scope.

So, what are the important topics in technology? As indicated above, when we think or hear the word "technology," we often think of computers or any type of device controlled by semiconductors that produces immediate responses from keyed or haptic commands. The immediacy and "knowledge" of these devices is based exclusively on code. Writing about technology is not and should not be equivalent to "technical writing." At the same time, it is also different from writing a novel or poetry. Technology, from the standpoint of mechanical, electronic, and microchip-based devices, has changed exponentially since the Second World War. So how is it possible, then, to write critically about technology, yet without exposing readers to too much jargon-laden terminologies? All too often, experts in their field use jargon that they think both experts and laypersons understand. In some cases, experts are unable to explain what they wish or intend to write about without the use of "jargon." At this point, one might wonder if these technology writers really understood what they were actually writing about. This can be quite a challenge when the goal is to write for a general audience. As technology progressed during the early, middle, and late 20th century, the number of technology writers were few in number. Today, the number of writers skyrocketed. But still, technology writing, much of it journalistic, would often be pedantic and jargon laden to the point that the layperson would not understand or keep up to date with contemporary issues in technology. To avert this problem, the following list considers recommendations for clarity and succinctness when writing about technology:

- Use concrete and straightforward terminology. Laypersons will understand the topic when the writer uses concrete, tangible words that connect ideas.
- The use of examples is essential when writing about technology because much of technology is all about novel ways of doing things in everyday life.
- Use contextual information that readers can relate to. Providing history or background knowledge will allow readers to place your writing into a context that they will understand.
- Implement illustrations if at all possible. We made this very same point in the previous chapter when discussing how brainstorming can be enhanced through the use of sequenced diagrams or figures. While diagrams may not necessarily tell the whole story, they do serve as an excellent ancillary to your text, and, if produced at the beginning stages of the writing process, they can, indeed, help the technology writer formulate ideas that will generate words on paper.

The following are suggested steps when beginning to research topics prior to writing about technology:

1. When thinking about issues in technology, identify issues that are contemporary and newsworthy. In the next section, I have provided topics that might be of interest from the point of view of technology. For an exploratory paper, you will need a topic that has three or more perspectives for investigation. If you want to write a position paper based on an argument that you believe needs to be addressed, you will use your answer to the question as your thesis statement.
2. Understand your writing assignment or the writing project you wish to pursue. What kind of research paper do you intend to write? Re-read your assignment sheet and any information in the textbook. Choose a technology topic that will help you research three or more perspectives on a specific issue in technology.

3. Read about the Topic. Once you find a topic of interest, begin to research the topic by finding seminal articles in the library or online using Google Scholar or a similar database. In terms of writing style, examine science magazines for non-technical audiences that address breaking news issues and technology research.
4. Examine credible sources and determine which questions can be answered with the relevant information that is available.
5. Write down keywords for any remaining questions, which will become the basis of the topic about which you are inquiring and researching. The goal is to narrow the number of issues to two, but no more than three.
6. While reviewing these issues, try to select the one that seems pertinent to contemporary societal problems.
7. At this point, start your preliminary research, and make modifications to your selected topic if needed. Use websites and databases to find good sources. One suggestion is to follow the links in hyperlinked articles that are written for non-specialists that can direct you to the original sources and research articles.

Now that we have covered the general steps and suggestions in good critical technology writing, we can now consider some perennial technology topics that have seemed contemporary in every generation. The following is a list of 31 questions that can help the critical writer of technology contemplate further exploration:

1. What are the long-term effects of living in a technological world? Are these mostly negative or positive?
2. Are children under 10 now growing up in a different world than college-age students did? How is it different, and what does that mean for them?
3. What is the most important new technology for solving world problems?
4. How has social media helped solve and create problems in countries outside the U.S.?
5. Will totalitarian governments continue to be able to control citizens' access to the Internet and social media?
6. How do social media, texting, cell phones, and the Internet make the world bigger? Smaller?

7. What are the implications of ever-increasing globalization through technology to the global economy?
8. Technology is changing so quickly that we are frequently using computers, software programs, and other technologies that have frustrating glitches and problems. Is there a solution?
9. How does our experience of social interactions with other humans influence the ways in which we interact with machines?
10. When does it become morally wrong to genetically engineer your child?
11. How have COVID-19 shutdowns, virtual school, and remote work changed our relationship to technology?
12. How is online education going to change the way students learn?
13. Does the Internet need controls or censorship? If so, what kind?
14. Do digital tools make us more or less productive at work?
15. To what extent is the development of new technologies having a negative effect?
16. How will technology change our lives in 20 years?
17. Should people get identity chips implanted under their skin?
18. Should people in all countries have equal access to technological developments?
19. Can video gaming help solve world problems?
20. How are brains different from computers?
21. Is organic food better for you than genetically modified foods?
22. What are genetically modified food technologies able to do? How does this compare with traditional plant breeding methods?
23. Should genetically modified food technologies be used to solve hunger issues?
24. Since it is now possible to sequence human genes to find out about possible future health risks, is that something everyone should have done? What are the advantages or disadvantages?
25. If people are subject to genetic testing, who has the right to that information? Should health-care companies and employers have access to that information?
26. If parents have genetic information about their children, when and how should they share it with the child?
27. What sort of genetic information should parents seek about their children and how might this influence raising that child?
28. Are self-driving cars a good or bad idea?
29. How might travel in the future be different?

30. Should information technologies and Internet availability make work from home the norm?
31. To what extent should corporations have access to personal and confidential information of public and non-public school students from preschool through the twelfth grade?

Conclusion

We began this chapter by showing the close association between the meanings of "technology" and "engineering." Many engineers think of these terms as one and of the same thing. That being said, a subset of writings on technological subjects consists of characteristics of engineering. At this point, you should have a good idea as to the types of issues that have arisen in the last decade or so in technological topics. From these examples, it is possible to consider further questions in the following categories: health technologies; genetic engineering technologies; human identity issues related to technology; technology in relationships and the impact of social media; military technology, such as drone warfare and nuclear weapons; information communication technologies (ICT); and computer science and coding—topics related to software development, machine learning, and programming apps. Further readings on the subject of adolescent development in technology writing include Cope, Kalantzis, and Abrams (2017) and Turner, Abrams, Katíc, and Donovan (2014). In the next chapter, we move beyond so-called silicon-based technologies and discover ways to write in the technological fields of engineering and architecture.

Note

1 In general, a heuristic is a mental pathway that is often used to simplify problems, challenges, or uncertain situations as a means of avoiding cognitive overload. Heuristics is not a novel phenomenon. They represent, in part, how the human brain evolved and has become wired, thus allowing individuals to reach reasonable inferences or solutions to complex problems in relatively short time frames.

· 6 ·

WRITING CRITICALLY IN ENGINEERING, ART, AND ARCHITECTURE

Picture it. The blueprint! You're an engineer who is about to consult a building panel of associates before the construction of a skyscraper is about to begin. You feel at home with the blueprint because you're a picture person. You might also be a mathematics person or a science person or a technology person, too. You have a passion for figures, diagrams, and the like. But one thing that you probably have a fixed mindset about is language use.

As engineers, architects, and others whose profession entails a love for sketching and drawing advance in their careers, they often become less involved with hands-on activity and verbal, onsite direction and more engaged in tasks that necessitate writing as a form of communication. The formats of writing in applied STEAM disciplines like engineering and architecture can be as basic as an email, as practical as a report or general specifications for the client or colleague, or as intense as a research article oriented to engineering or architectural research.

Typically, unlike writing in the natural sciences and mathematics, writing in engineering or architecture requires the professional to consider multiple genres of writing. While the biologist or mathematician writes research articles, reports, or books, the engineer and architect frequently convey their ideas through both formal, scientific and everyday, spontaneous forms of communication (Paretti, Eriksson, & Gustafsson, 2019). Note the many similarities between writing in engineering and architecture and that of the other STEAM disciplines. For one thing, succinct, clear, and concise prose is perhaps the biggest takeaway when it comes to writing in STEAM. It's no different for architects and engineers.

While natural scientists and mathematicians will write formal articles and books, their informal writing is often in the form of practitioner manuals, textbooks, or articles for STEM teachers. However, this is not necessarily so for the architect or engineer; for them, there are numerous modes of communication in which they are required to engage professionally—whether formally or informally. The following are some of these genres: emails, letters, reports, business cases, inspection reports, specifications and instructions, and academic writing, which includes dissertation theses, articles, and books. What follows are accounts of each of these genres as forms of written communication for architects and engineers. Following the first of these genres, email writing, are descriptions of understanding proper written communication protocol and technique of various features of informal and formal writing that will enable architects and engineers with the tools needed to compose strong and effective written forms of interaction within their fields. Note that, unless additional features are described for genres other than email communications, descriptions of features shared within the section on email correspondence are applicable to the other genres as well.

Email Writing

Unlike professionals in other STEAM areas, architects and engineers are not often teachers or researchers. Therefore, email use is essential in getting information to a client or business associate as speedy as possible. Indeed, emails have several advantages; time and cost saving, convenience (compare to telephone and fax use and wiring funds and information), ability to forward to or include other individuals, international communication, and read on screens are just some of the many benefits to email communication. Despite these advantages, however, emails, given that they are seen as written documents, can be used in a court of law. In other words, an email can be used against the sender for any language that is either considered or interpreted as offensive or hateful in tone or in writing. In most cases, as soon as one clicks the "Send" button, that's it; it's sent. There is no way to retract the email in the vast ether called the internet. The email will end up in the recipient's inbox. Although you can delete it from your sent box, it cannot be undone. It is now a form of evidence. Email senders, then, must realize the importance of checking their writing for tone, content, grammar and syntax, greeting the recipient, and overall organization.

Tone

To begin, tone must be considered when writing an email. As humans, it is natural to write quickly as if we were speaking to the recipient. At times, our writing can seem informal, overly personal, or aggressive to the recipient, and the fact that our email is received at lightning speed all the more confirms this impression. Unfortunately, due to the fact that the email sender is not in the same room, or not in the same building, city, state, or country for that matter, the recipient cannot detect our facial expressions, gestures, and tone of voice, and while emojis or emoticons might tend to quell the recipient's negative interpretations of the writer's unintended (or intended) language, emojis, in and of themselves, are inappropriate for formal communication. Recipients only go by your written words, so it's important for the email writer to anticipate what the reader will think and how the reader will interpret the language within the email. Also, although one might be friendly with a colleague, business or academic emails must be formal since the friendly colleague may inadvertently forward the email to numerous individuals or include as part of an email chain. In short, note that emails are not necessarily limited to the intended recipient as they can be passed along to others without the knowledge of the sender.

The upshot, then, is that email writing requires a good deal of thought and preparation. It is not a substitute for verbalized speech. If emotions take over rational thought when composing an email, take a break or a walk as a way to calm down and to lessen the extent of negative emotion. If the writer is upset with the recipient and hastily sends the email, the recipient will undoubtedly interpret the tone as anything from rude or disrespectful to threatening. When feeling irritable toward a client or colleague, wait some time before the email is sent so that the relationship does not go from disagreeable to threatening or intimidating. At the very least, type the email address as the last step so that there is time to rethink what was written.

Grammar and Mechanics

Grammar represents the assembly of both spoken and written language. It is about taking words within different parts of speech and combining them in a way that form meaningful sentences. In short, grammar is the structure of language that governs semantics—how ideas in the form of speaking and writing make meaning to listeners. People in past (and some present) societies that have had limited literacy ability depended on speech as a form of communicating ideas. With extremely few exceptions, grammar comes naturally to humans within the first few years after birth. In other words, developing toddlers and young children don't have to learn formal grammatical rules in order to string words together for listeners to understand what they are saying. In contrast, young people will almost certainly need to learn grammar formally when developing their formal writing skills. Mechanics refers to the rules of written language. These rules include spelling, capitalization, and punctuation. In tandem, good grammar and mechanics form the foundation of written communication.

When writing an email, or any correspondence document for that matter, proper grammar and mechanics are important and must be taken into account. There are several reasons for this. First, clarity is of utmost importance. Few, if any, recipients will want to read correspondence that possesses poor grammar. Such writing is, more often than not, difficult to understand, and, therefore, may stifle further correspondence. Good grammar and mechanics skills allow for easy-to-understand communication between writer and reader. Second, it is important to maintain rapport with others. Writers with poor grammar and mechanics skills run the risk of losing clients, many of whom would want their services completed not only by skillful architects or engineers, but professionals who can communicate ideas clearly and concisely. Third, good grammar and mechanics can help writers avoid ambiguity. Take the following sentence as an example: "Before eating lunch, the blueprint was not yet finished." This sentence includes a dangling modifier: it is unclear as to who (or what) is doing the eating. Clearly, blueprints don't eat. One way to fix this sentence is to write "Before we ate lunch, the blueprint was not yet finished." Fourth, sometimes we will need to communicate with people from other countries or from those whose native language is not our own. One side will need to understand the other side's language. So, when someone is attempting to read in a language that is not their native language, the text must have good grammar and mechanics. Fifth, good grammar and mechanics demonstrates respect toward your readers. One who makes these types of mistakes in written correspondence or other documents will give the impression to the readers that they are not important enough to take the writer seriously. Sixth, good grammar and mechanics means that one's message is being conveyed. This will help the writer because the audience will understand what is being expressed so that communication will be successful. In addition, poor grammar and mechanics makes more work for an editor—especially for formal writing, such as articles and book chapters. The editor's job is to check writing for organization, argument, and suitability, not to fix errors or rewrite passages.

The following suggestions might help STEAM writers improve their grammar and mechanics skills. First, make a list of the most common errors. Second, read written communication aloud, or listen to someone else read what was written. Doing so will help the writer determine what "sounds right" and what doesn't. Next, do not depend solely on the spelling and grammar check tool in your word processing software. Oftentimes, these tools simply make suggestions that may not always provide the correct solution. For example, examine this sentence: "The engineer rights clearly." This sentence implies that the engineer does a good job making something right or appropriate—a meaning that is demonstrably weak in this context because the use of "right" as a verb is somewhat awkward. Instead, the author probably meant that the engineer knows how to compose (i.e., write) a letter clearly. The spelling or grammar tool will not pick up this error. This issue makes evident the need for writers to proofread anything that they finish writing. There should be no exception to this rule. Lastly, learn more about your problem areas in grammar and mechanics. For example, those who write a lot of dangling modifiers or who often write run-on sentences would want to see more examples and ways to fix these problems.

Content

While email writing is quite amenable to size, it is necessary to consider the extent to which the length of the email can be favorable or detrimental to the intentions of the writer. For example, an engineer or architect can include a technical report in the body of an email, along with charts and diagrams. However, the formatting and overall structure of the report can be altered during transmission of the email. Therefore, it might be more fruitful to include the report as an attachment, which will reduce the word count of your email drastically, thus making the text in the body of the email succinct. Shorter texts in the email can instruct the reader about the content of the attached documents, which will make for easier and clearer reading.

Greeting the Recipient

When communicating professionally—with clients, superiors, employees (most of the time), or outside affiliates—the salutation is just as important as much of the body of the email. It would be a mistake to think that the way you greet a client or another professional is not necessary. If you offend your email reader, that person may not read any further—especially if your recipient is a client. If the recipient is a colleague, that person's opinion of you may be affected.

Indeed, there are salutations to avoid when communicating through work emails. As for salutation, I live by one cardinal rule that, unfortunately, I have not seen in any of the writing-in-discipline literature. And that rule is to precede the name of the individual (or individuals) with a word of greeting. All too often, I have been the recipient of emails from professionals who write a salutation without a greeting word: "Mr. Ness," "Dr. Ness," "Daniel," and the like. Recipients of emails that have these salutations might find the sender, especially if the sender is not very familiar, abrupt and may feel that the sender is about to reprimand them. I know of numerous people who might not mind on the surface that senders write to them solely with the recipient's name in the salutation.

Another poor salutation is one that includes solely the greeting: "Hi," "Hello," or even more egregious, "Hey." Not only are these salutations too casual; they also reek of insensitivity. People like to be called by their name. If the person is an associate, "Hi [First name]" or "Hello [First name]" works quite well. If the individual is a client, a distant superior, or an external associate, "Hi [Ms. Last Name]," "Dear [Mr. Last Name]," "Good morning [Mx. Last Name]," or "Hello [Dr. Last Name]" are appropriate.

Proofreading or Checking

Individuals who are preparing to become or are currently engineers or architects will undoubtedly check hard copy business letters and reports for accuracy. But, oftentimes, they will send emails without proofreading or checking for either unforeseen errors or initial tone, which may seem harsh or abrupt to the recipient. There are two general reasons to fastidiously check one's email: to avoid inadvertent errors and to verify to whom the email is addressed.

All too often, emails are unchecked with errors and run the risk of being read by colleagues, senior administrators, or clients. Indeed, errors lead to incorrect information, which questions the integrity of the email's author. Furthermore, proofreading and checking also involves who the recipient is, and, if more than one person will be emailed or copied, who the recipients are. Therefore, unchecked emails can jeopardize the author's reputation if the email is sent accidentally to the wrong individual or group.

Just as important as to whom one is writing an email is to whom one responds with regard to an email addressed to the writer. There have been many a time when I would be the recipient of an original email sent to a long list of copied recipients only to be on the receiving end of a follow-up email from one of the copied individuals who either decided to, or inadvertently, responded by clicking the "reply all" button. Many emails with long lists of recipients usually involve future events or actions on the part of the recipient and the only one who needs to know about your intentions is the initial sender. The upshot is this: Don't reply to all of the recipients. Aside from the potential embarrassment of replying to all (i.e., recipients can see who is sending the response to everyone), no one needs to know one's intentions in response to the initial call to action or an upcoming event. Moreover, a recipient can rightly infer that the responder is replying all for self-absorbed, egocentric reasons, for example, that the responder wants the entire group to know her, his, or their intentions.

Texting Language

Avoid using texting language, such as text abbreviations and acronyms—that is, popular language and spelling of words or phrases, used when texting others on mobile phones and computers. In texting, words or phrases are most often condensed into a series of one to six or seven letters, numbers, or combinations of letters, numbers, and symbols. Many reasons for avoiding this language account for this. To begin, texting language tends to be overly informal because we tend to text those who are closest to us—family members, friends, significant others, close colleagues, and acquaintances of friends are just a few of these groups. One will almost never find texting language in formal emails and other documents discussed in this chapter or in writing genres in the other chapters. Second, the use of text abbreviations shows a lack of professional etiquette and protocol among the vast majority of professionals. For better or worse, society still embraces the use of official languages—that is, the dominant languages spoken in individual countries. So, so-called Standard English, French, traditionally written Chinese, German, Spanish, and the many other official languages of the world must be used—whether in email or letter form. In addition, texting language can be unclear and misunderstood by the reader. Take the following example and see if you can decipher it: "TYSM 4 ur letter. I'll!= b aO 4 3 dAz. I will call u 2mro 2 plan 4 a d8 2 meet. U can tL every1 about it. cul8r n hand! D." This short letter is obviously hard to decipher for some people and clearly inappropriate for semi-formal or formal emails or letters. In case you're wondering, the note reads as follows: "Thank you so much for your letter. I will not be around for three days. I will call you tomorrow to plan for a date to meet. You can tell everyone about it. I will see you later and have a nice day! Daniel"

Letters

For the most part, emails as a form of communication with colleagues and clients in engineering or architecture, and STEAM in general, has replaced letter writing (i.e., paper and pen or typewritten documents). However, there are still occasions when engineers or architects need to compose and send hard copy letters or facsimiles (i.e., faxes), and many of these documents are in the form of legal or business correspondence. Sometimes the letter writer needs to address issues of appointment, promotion, or modifications of working conditions—formal topics that are best in the form of an official, hard copy letter (McCall, Fillenwarth, & Berdanier, 2020).

Like emails, letters must have appropriate salutations. A salutation should begin with "Dear"; however, "Dear Sir" or "Dear Madam" are now antiquated titles. Moreover, with respect to each person's identity, writers must consider to whom the letter is addressed. Female addressees should go by Ms., and males by Mr. Those who refer to themselves as non-binary or who want their identities to be gender neutral should be addressed with the honorific Mx., which is generally pronounced "mix." It is also the case that the recipient received a doctorate and should therefore be addressed "Dr."

Van Emden and Becker (2018) point out the importance of writing as simplistic and clear as possible. It is common, for example, for the writer to use ambiguous, tautological, or wordy language that can otherwise be summed up more succinctly. Let's take the following example: "I am writing in response to your inquiry that I received on February 27th." First, avoid beginning sentences with "I," especially as the first word of the entire letter. Doing so lessens the perception of self-aggrandizement. Sentences that begin with "We" are better since this shows a sense of working cooperatively. Those that begin with "You" is also better as this allows the writer to structure the opening of the letter in a way that shows the reader's importance. Second, avoid the phrase "am writing," given the fact that the recipient is reading the letter and obviously knows the writer wrote it. Third, "that I received on" is superfluous in that you are clearly responding to the inquiry. The following sentence demonstrates a better way to open the letter: "Thank you for your inquiry on February 27." Then the writer can get into addressing the issues within the body of the inquiry.

Business Case Documents

A business case document is a formally written composition that is similar to a proposal in that it forms the basis of a new project or the development of a new product. Due to its overall complexity, the business case is time-consuming and may require the writer to engage in a good deal of investigation and preparation. The reason why business cases are complex is that it takes time for the engineer or architect to identify the need of a project or new product and to obtain permission from authoritative entities, which can take many months to secure. The main sections of a business case are the Executive Summary, Current Situation, Options, Cost–Benefit Analysis, Timeline, and Conclusions which includes a recommendation.

The initial section of most business case documents is called the Executive Summary. This section is analogous to the abstract of a paper or technical report in that it is intended to convey the general idea of the entire document in condensed form. Also, like an abstract, a business case Executive Summary can be considered the most significant section of the document because it summarizes the entire plan. Because it encapsulates the entire project, the engineer or architect will write this section last but place it first in the document. After the Executive Summary comes the Current Situation, which outlines the present state of a lot or the lack of a needed product that the proposed project is supposed to address. The general idea is that the intended project will improve upon the current situation or solve a current problem. The writer should address the reasons why the project is needed and what would happen if the current situation were to continue.

Next comes a section on Options, which addresses not only the proposed project but other possibilities as well. The author will need to discuss strengths and weakness of each of the options and the different components that they cover. These would include cost, labor, and ethical validity. Doing so will demonstrate the author's ingenuity and impartiality. This section should end with the author's project recommendation and the reasons for its proposed implementation. After Options comes the Cost–Benefit Analysis, which shows the benefits and limitations regarding the project's cost. The cost–benefit argument should be based on past precedent and charts should be used for easy access to the numbers. All projects require a timeline, and this section follows the Cost–Benefit Analysis. The author should base the timeline on the time it takes to complete a given project as well as the time needed to garner materials. Given unforeseen circumstances, indicating more time than anticipated is always more desirable.

Technical Notes and Technical Reports

Technical notes are shorter than technical reports and convey information that deals with a new development or procedure that might need to be implemented for the purpose of efficiency or cost effectiveness. Technical notes usually take the form of articles in technical or practitioner-based journals and, like action research studies, they can be used within an organization for the purpose of improvement of a product or a technique. The overall outline of a technical note includes the abstract, followed by a literature review, methodology, results, and discussion. While technical notes start out as short, research-based papers, they can be developed into longer articles that can be published, and therefore circulated to a larger readership.

Like technical notes, technical reports for engineers and architects are essentially analogous to the natural or social science research paper. They are similar to their shorter cousins in that they contain a title page, abstract, introduction, literature review, methods, results, and discussion. The following shows the overall structure of the typical engineering or architecture-based technical report. Note that any specific report may have different headings and subheadings but what appears below is the basic format for all reports:

Table of Contents	
Abstract	i
Introduction	1
Methods	X
First Subheading (If Subheadings Exist)	X
Second Subheading (If Subheadings Exist)	X
Third Subheading (If One Exists)	X
Results	X
First Subheading (If Subheadings Exist)	X
Second Subheading (If Subheadings Exist)	X
Discussion	X
Conclusions	X
Appendix A: Title of Appendix A	X
Appendix B: Title of Appendix B	X
References	X

In parallel with science research papers that we discussed in Chapter 4, each laboratory study requires a report of some kind. This document provides guidelines for writing laboratory reports. Specific instructions for the report should be given with each exercise. Check with your instructor if any instructions are not clear. The report is an account of your work in the laboratory. There are two target audiences for reports: colleagues and laypersons. The first audience consists of people with a background similar to yours (e.g., other engineers, technicians, etc.) who want to be able to duplicate your results. For this purpose, we will use a formal report. The second audience consists of people who have some understanding of the process/problem and want a clear, concise presentation of the results (e.g., managers, other semi-interested engineers, etc.).

Title Page

The title page should contain the title of your report. The title should refer directly to the experiment or empirical study that is being conducted. It should also include the names of all laboratory writers and the date. All this information should be centered, near the top of the page.

Abstract

The abstract is a brief, single paragraph that describes the objective of the laboratory activity, the results obtained, and conclusions. Abstracts allow interested readers conducting research to reach informed decisions about looking further into the report. The abstract should appear on the page that follows the title page.

Introduction

The Introduction section provides a short description of the problem being addressed in the laboratory exercise. It is an explanation of how the objective of the laboratory is achieved and provides enough background for the reader to fully understand the rest of the report.

Method

The Method section describes the processes and steps needed to engage in the science study. It contains system diagrams, equations as appropriate, and an explanation of any equipment or measurement techniques needed. The Method section should be detailed enough so that someone could duplicate the results.

Results

The Results section explains what was found in the study and what, if any, tools were used to gather the data for analysis. The Results may have data tables, graphs, screen captures of specific displays, and narrative discussions. The interpretation and subsequent analysis of the data must be clear and unambiguous.

Discussion

The Discussion section should allow the reader to interpret the results of the study. If there is a particular conceptual insight revealed in the study, it should be discussed. If there were difficult or unusual aspects of system construction or measurement, they should be discussed in this section. Write this section in the context of the study's objective.

References

All reports must have a References page. This page should include all references that are cited within the body of the report. In addition, each reference must be written using a specified format. The primary format used by scientists is that which is associated with the Chicago Manual of Style. Research papers in the social sciences often adhere American Psychological Association (APA) formatting.

Appendices

Appendices can be useful and important additions to a technical report since they generally include information that might be tangential to the report but important in terms of explaining specific methodological approaches that would be too wordy or verbose to include in the general technical report narrative. The technical report should include appendices only if there is material that one can reference in the event the procedure involves written protocols, assessments, or evaluations. The writer should not assume that all reports contain appendices.

Equations

Equations are common in many technical reports. Reports will need to be prepared with word processing software such as Microsoft Word or Pages, and all equations should be prepared using Microsoft Equation Editor or similar software. When preparing text containing equations, the writer should follow an established set of criteria that most engineers and some architects follow. The following principles are useful when including equations in the technical report:

1. All variables are written in italics. For example, $\dfrac{df}{dt} = \lim\limits_{h \to 0} \dfrac{f(t+h) - f(t)}{h}$ can be included by using Microsoft Equation, LaTex commands, MathML elements, or MathType.
2. All units are written in Roman text (i.e., not italics) with a space between the number and the unit symbol. For example, six kilograms should be written as 6 kg.
3. Indeed, there are exceptions to the rule: Common functions such as the trigonometric functions are italicized, for example, $\cos\theta$ is acceptable.
4. Matrices and vectors are written in bold and in Roman font. For example, $\mathbf{v} = 6\hat{i} + 3\hat{j}$ would be appropriate. Fortunately, these equations

and mathematical terms are available in Microsoft Equation, LaTeX, MathML, and MathType.

Graphs

Like equations, graphs also are widely used in technical reports some of the criteria to follow with regard to graphs include the following:

1. All graphs must have axis labels and a title
2. If there are multiple plots on the same graph, a legend is essential.

Microsoft Excel makes adherence to these and other accepted graph preparation procedures very straightforward, and all graphs must be prepared using Microsoft Excel or similar software. Note: When a graph or any information is prepared in landscape format, the page must be oriented in the report so that the bottom of the page is to the reader's right-hand side.

Graphics

Graphics in your report, such as system diagrams, should be prepared using software with graphics capabilities. Examples of such software are PSpice, Microsoft Word, Adobe Photoshop, AutoCAD, and SolidWorks. Figures should be properly sized and easily readable. Avoid scanned images or images copied from another document, unless absolutely necessary. Use the "crop" function to eliminate unnecessary parts of figures and make them more adaptable in the document. When using a picture from another document, it will be necessary to cite its source.

Other Reports and Specification Documents

Inspection reports represent an additional type of writing genre for architects and engineers. These reports are easier to write because they are most often presented in the form of standard documents that have their own templates. Professionals, then, can easily write the answers to each of the detailed items within their prospective line spaces. In addition to inspection reports, architects and engineers often need to complete specification or instruction documents.

Academic Writing

For architects or engineers who are planning to write research articles, theses, dissertations, books, or book chapters, I would refer readers to Chapter 4 or Chapter 7, which discuss academic writing in more detail. In general, however, the overall structure of an academic paper in engineering and architecture is similar to those in the natural and social sciences (Spector & Damron, 2017). The author engages in academic writing by starting with a research question, which serves as the catalyst for investigation. The research question provides the researcher with the opportunity to examine the research literature in the areas that are germane to the answer to the research question. Oftentimes, the literature review is divided by themes—areas of interest that are directly related to the overall thesis. Let's take the following research topic as an example: A dissertation entitled *Landscape Infrastructure: Urbanism beyond Engineering*, by Pierre Bélanger (2013). The author states in the abstract, "As ecology becomes the new engineering, the project of Landscape Infrastructure—a contemporary, synthetic alignment of the disciplines of landscape architecture, economics and cost-benefit analysis, civil engineering, and urban planning—is proposed here" (p. 9). Based on the abstract, and even the title, one can infer several prominent themes connected with this dissertation. These include, but are not limited to, urban planning, climate change, population mobility, resource flows, landscape architecture, and civil engineering. The author's job, then, is to identify publications that support these strands of knowledge. In doing so, the author is acknowledging these important themes and demonstrating to the reader how the identified supporting publications help to bolster the research claim and show the existence of a gap in the literature that he is attempting to fill. In short, the literature review includes the themes of research that one examines to help support the thesis of the paper or dissertation.

The literature review is then followed by a restatement of the purpose, which shows the need for a gap to be potentially filled. As illustrated in other STEAM domains, academic papers continue with a Methods section that identifies the procedures that the author will follow to address the problem—i.e., the research question(s). Following the Methods is the Results section, where the reader will learn the author's findings. If the author uses quantitative methodological approaches, the Results will undoubtedly include multiple statistical procedures to arrive at these findings. The author interprets these findings using a combination of writing procedures that include mathematical formulas, usually shown pictorially in charts, and explanation and analysis in prose.

The author then interprets the analytical justification in the Discussion, the last main section of the academic research paper. The author's interpretation is intended to accomplish at least two things: 1) it should tell the reader how the author addressed or filled the gap in the research; and 2) it should demonstrate to the reader how the author's thesis addresses the overall need of the study—to put it in everyday terms, how the identified answer to the problem makes society better.

In the following chapter, we will recognize and appreciate the interconnections among scientific, technological, engineering, and architectural writing, and how they are related to writing in mathematics.

Questions

1. Outline the information that is needed in the following excerpt for a business case: You are an architect in an organization and, due to an increase in projects, you found that your workload is increasing and realize that the surge in jobs requires you to seek part-time assistance. Make your case to your employer and include as much supporting information as possible to do so.
2. Identify the main stylistic problem in the following excerpt:

> Mr. Smith!
> It has come to my attention that your electric vehicles are not as technologically cutting-edge as they could be. Microprocessors are more than just a popular technological catchphrase; they are something that can be effortlessly implemented into existing vehicles and will add countless dimensions to their capabilities . . . These are of course miniscule examples in a sea of things that can be accomplished with microprocessors. There are much more useful and innovative things that could be done to improve both the mechanical and ergonomic aspects, which would put you light-years ahead of your closest competitors, all the while fattening your pockets . . . I enthusiastically look forward to meeting with you!

Each of the following excerpts has an ambiguity, which is a word or phrase with more than one meaning. While ambiguities are acceptable among authors in the humanities, engineers and scientists can be litigated for ambiguities. In the following sentences, identify the source of the ambiguity: (1) improper syntax (word order), (2) missing comma, (3) unclear pronoun reference, or (4) grouping of conflicting words.

3. We propose to provide the above engineering services hourly based on the following estimates.
4. Compared with the pollution of the average coal-fired plant, the thermal pollution of a nuclear power plant is less than 2% more.
5. As airplane designs change the anti-ice systems also have to change.
6. At this time, the Department of Energy is only considering Yucca Mountain as a possible storage site for nuclear waste. Other possible sites are excluded from discussion.
7. If the airplane waits too long to take off the de-ice fluid can dissipate.
8. The beams are placed with respect to the posts so that while one beam passes the opposite beam is completely blocked.
9. The structural engineer who arrived to survey the rubble and identify the reason for the collapse found that the disaster was caused by the failure of the hanger bolts which bound the roof's steel trusses to its hangers and were worn out as the structure moved in the wind.
10. Avoiding complicated multi-ordered calculations, the equations come from fundamental definitions of mass flow, work, and efficiency.
11. To provide spill protection, all tanks were equipped with basins and automatic shutoff devices or overfill alarms or ball float valves.
12. The O-rings cannot expand in 32-degree weather and the gas finds spaces in the joints, which led to the explosion of the booster and then the shuttle itself.

· 7 ·
WRITING CRITICALLY IN MATHEMATICS

Writing in, on, or about mathematics is a venture of expressing quantitatively-based concepts; it's a way to convey thoughts that enable possible ways of solving everyday or spontaneous problems, which includes, but is certainly not limited to, issues related to compound interest, how mortgages work, getting out of credit card debt, costs of commuting to work or school, understanding health data, the mathematics of limiting carbon emissions, increasing knowledge about health and life insurance, optimizing your diet, making great estimates using few data, proving financial fraud, and, yes, even exposing cheating in elections. It's also a way to express ideas related to theoretical issues, such as deducing proofs, creating and developing theorems or corollaries, or engaging in mathematics history (Miller, 2015).

Reading and Reviewing Mathematics Content

It might be useful to know that engaging in mathematical writing more than likely is not the same as writing novels, essays, letters, or short stories. Likewise, it is not what people do when they write experimental or non-mathematical papers or reports. Rather, there are both general and specific overarching themes that apply to mathematical writing that do not necessarily apply to other disciplines. We will begin first by encountering these themes. We will then apply them to the writing stage—putting ideas into words, sentences, and larger writing constructs.

What Is Different About Writing in Mathematics?

We have discussed many similarities between writing in STEAM and non-STEAM disciplines. Language used in subjects like history, English, philosophy, or sociology is rich in that it allows for ambiguity and nuance. For example, a passage from writing in history will often use words that convey indefinite characteristics of a given event. Because the narration of historical events cannot be reproduced as it can be in the natural sciences or in mathematics, evidence used to support the existence of a historical figure or occurrence of a past event cannot demonstrate certainty; rather, historical evidence is based on corroborating sources that only increase the probability that a person existed, or an event took place. Not so in mathematics: Mathematical language is concise and must be unequivocal and explicit in its terminology (Halmos, 1970; Krantz, 1997). Mathematics writing requires slow reading, especially if new ideas are being expressed. Therefore, it often must be read and pondered several times.

I like to stress what Anderson (1990) refers to as class structuring when he teaches mathematics. In Anderson's words, "I facilitate the creation of study groups of no more than three or four people of the students' choice. These groups allow collective study ... and in-class group work. The study groups are responsible for one or two 'progress reports' (others may call them 'tests' or 'exams', but I prefer my term because I want to convey the message that I am optimistic about my students' performance and concerned primarily with building mathematical confidence)" (p. 357).

Overarching Themes in Mathematical Writing to Think About

There are several ways to write mathematics, especially when using abstract symbolism that will enable readers to understand what the author is attempting to express (Knuth et al., 1989). In his online pdf, Bertsekas (2002), a faculty of engineering at MIT, outlines several rules to consider when writing in mathematics. The first thing the writer of math should consider is to organize writing in segments.

Various genres have their own basic units. For example, the short story has the paragraph as its fundamental unit; the film or play has the scene; the PowerPoint presentation has the slide; and the opera has the number or aria or recitative as its fundamental unit. It follows, then, to ask the seminal question: What is the fundamental unit of writing in mathematics papers or publications? Writers of mathematics call this unit a segment, which is an entity that remains clear, concise, and unconvoluted in its entirety. Examples of segments include, but are not limited to, a mathematical result and its proof, a mathematical example, an appendix, an abstract, and even a paragraph—especially when the writer of mathematics is attempting to convey a point of argumentation or contradiction in another's publication or presentation. Like a paragraph in an essay, a segment should function on its own. Figure 7.1 shows the general layout and structure of a segment.

The second thing the writer of math should consider is to write segments in a linear fashion. In general, math arguments or positions should be placed near where they are used and not in some other location. It follows, then, that definitions and other mathematical components should be placed close to where they are used. For example, when considering order of influence or optimization, make it easier for the reader to construe what is being presented. Figure 7.2 demonstrates the flow of specific arguments with both linear and non-linear versions of initial level and subsequent level arguments.

Third, when writing (or reading) mathematics, it is acceptable to consider hierarchical development. So, arguments or results used repeatedly may be placed in special segments for efficiency. See Figure 7.3. Special segments, then, can be created for both mathematical and historical material, such as mathematics background and notation.

Fourth, writers of mathematics should be consistent not only in terms of form but also in notation. Examples of consistency in notation include, but are not limited to, the use of upper-case letters for random variables and lower-case for values; or subscripts for sequences and superscripts for exponents. Further along this line of thinking, it is suggested for the writer of mathematics to use suggestive, mnemonic notation. For example, we can use "S" for set, f for function, t for time, and so forth. In short, when writing mathematics, avoid unnecessary notation.

Fifth, state results consistently. In other words, keep your language and format simple and consistent. Keep distractions to a minimum by making the interesting content you have to offer stand out. To do this, use similar formats in similar situations. For example, the first set of propositions represents a bad set of propositions, and the second set represents a good one:

Proposition 1: If A and B hold, then C and D hold.
Proposition 2: C' and D' hold, assuming that A' and B' are true.

Instead, think of rewording this set of propositions in a parallel manner as in the following:

Proposition 1: If A and B hold, then C and D hold.
Proposition 2: If A' and B' hold, then C' and D' hold.

Begin each segment with a short introduction and overview of what the reader should expect. That is, keep the reader informed about where you are and where you are planning to go with your mathematics thesis. Stringing together seemingly aimless statements will not help the reader understand your intentions. For example, you can write "It turns out, then, that there are five chickens and three cows. To see this, examine the two algebraic statements I used to arrive at my answer..."

Next, when referencing, it is strongly suggested that the writer of mathematics use suggestive references. In other words, refrain from using numbered references when referring to specific equations or concepts because this adds extra time for the reader and can possibly break concentration. Instead, refer to equations, concepts, or results by name. It is safe to say, then, that repetition, in moderation, is not a bad thing.

Next, the mathematics writer should consider examples and counterexamples when critiquing other mathematics writing. A simple example may have enormous value when explaining a mathematical situation. Also, the mathematics writer is encouraged to use counterexamples as a means of clarifying the limitations of the analysis and the need to provide mathematical assumptions.

And lastly, it is important for the mathematics writer to demonstrate ideas using pictures and other forms of visualization. Doing so allows the reader to make connections among the different ideas that you are presenting. To this end, it will be important that figures and diagrams are kept simple and uncluttered. Moreover, captions should be used to reinforce the text and not simply repeat it. Similarly, the figure should illustrate the main idea of your argument without constraint.

Putting Ideas into Words

Now that we've encountered the overarching themes to writing in mathematics, we are now at the point where we can begin to put our mathematical ideas into words. We start by identifying what children do when they initially write in or about mathematics.

The development of mathematical writing is a murky enterprise because it is difficult to know the extent to which any young child is familiarized and experienced with the writing of mathematical symbolism. Children have their unique experiences that underscore not only how but when they will go about writing in or about mathematics. Therefore, we will need to look briefly to the area of developmental and cognitive psychology for some answers about how mathematical writing unfolds from the early years to adolescence.

To begin, we need to clear up a basic assumption: Schooled children, or children overall, do not learn how to write in mathematics at the same time, the same pace, or even with the same technical level. Again, this has almost everything to do with the level of experiences, or lack thereof, in the child's life that contribute to the habit of writing. In just about the same way that young children learn how to write the letters of the alphabet to form words, then sentences, then paragraphs, and essays, so, too, do young children need the experience to learn how to write the numerals from 0 to 9, operational symbols (i.e., $+, -, \times,$ and $\div$), symbols for equality (i.e., $=, <, >,$ and eventually $\neq, \leq, \geq, \cong,$ and $\approx$), symbols for fractions, ratios, and proportions, as well as symbols needed when encountering algebra, geometry, statistics, calculus, and more advanced areas of mathematics. When starting mathematical writing, children's written (or spoken) cognitive errors in mathematics often have sensible origins (Ginsburg, 1989). For example, if one were to ask a Kindergartner or first grader to write the number "twenty-one," the child might erroneously write "201" because they heard the spoken words "twenty" and "one." Thus, they often string the numerals together. In another example involving operations, a young child might conclude that $\frac{2\ \ 1}{2\ \ 3} - 4$. While an older child or adult might find this conclusion erroneous, it still has sensible origins—after all, $4 - 1 = 3$.

Accordingly, like the beginning stages of writing prose (words, sentences, paragraphs, essays, and larger blocks of writing), learning to write in mathematics involves learning how to form numerals, learning the spellings of the speech sounds of numbers, developing legible handwriting, and learning to select the best ways to convey meaning in mathematical terms. Learning to write in mathematics necessitates the child to develop an understanding of the conventions not only of mathematical symbols but also mathematical terms, statements, equations, and proofs, just as the young student learns the conventions for sentence and paragraph structure, punctuation, capitalization, and organization of longer pieces of writing. As children develop and hone their writing skills, they learn the purposes for writing and the special requirements of each, and they experience the processes of writing (e.g., planning, drafting, revising).

As teachers, we often take for granted what others, younger and older, should know, and writing development in mathematics or in something else is no exception (Fuehrer, 2009). It requires several steps, but as stated above, the steps are not innate, instinctual, or based on stimulus-response protocols; children from the early years through adolescence require instruction and practice. Moreover, it should not be considered a task; rather, the young student should be drawn to the activity of writing and not be compelled to learn. Not everyone learns to read and write in the same way or by following the same sequence. Teachers vary the procedures they use to develop both non-STEAM and STEAM literacy in their students depending on the teachers' experience, knowledge, pedagogical philosophies, the availability of materials and time, and, unfortunately, the dictates from those higher in command within education institutions as well as others outside the classroom.

Writing When Problem-Solving

We now encounter the critical writing of Angela, a 20-year-old student of Latinx background. In examining the following written dialogue, note that the mathematics writing involves both mathematical concepts as well as written explanations in prose form. So, in a particular assignment in my mathematics methods course, I posed the following situation to Angela and the other mathematics students in class:

> Ask one elementary and one secondary school student to solve the following problem (If you are unable to find an elementary school student, you may ask an older student or adult). Please feel free to ask more than two students or also to give the problem to

an older student, friend, neighbor, parent, or grandparent. Your role is not to teach or assist, but to observe and analyze how the students solved the following problem! "One day, Farmer Jones was counting her cows and chickens. She noticed 22 legs and that there were 8 animals in all. How many of each kind of animal (cows and chickens) did she have?" Since we all will share the results from these students during a class session, be prepared to talk about the way your learners approached the problem, the mathematics the student utilized to solve the problem, any misconception the student has, and what questions you might pose for this student to move him/her to a new level of understanding. Please bring students' writing/drawings involved in solving this problem. Their attempted solutions will be the basis for class discussion. Please write a reflection in which you will share insights gained as a result of your participant responses.

The answers I received from my pre-service teacher students were remarkably unique and demonstrated the expansive gamut of possible findings that one could think of—and not just mathematical findings! For example, some of my college students asked both elementary and secondary students whose responses demonstrated critical inquiry from both a physical and social standpoint. More specifically, many of the non-mathematical critiques of the children were those that indicated how specious this particular problem was. For example, one elementary student said, "How can this be real? The heads and legs of cows and chickens are so different!" Yet another said, "Cows are so much taller than chickens! So, this isn't really a problem." Other K-12 students, however, focused their attention on social constructs. For example, one secondary student said, "There's no way this can be called a problem. With all the real difficult, and sometimes dangerous, work they sometimes have to do, what farmer in their right mind would ever have the time, or the luxury, to consider this a 'problem' that they have on their hands . . ." One fourth grader said, "I'm glad that Farmer Jones is a she . . ." Yet another student, a seventh grader, questioned why the name chosen for this math problem is not necessarily representative of the surnames of the overall US population.

Before examining Angela's findings as she writes critically about the mathematical thinking processes of two K-12 students, I emphasize that the distinctiveness of this type of question, regardless of its alleged speciousness, is the fact that critical writing in mathematics can be seen in a layered manner—namely, we see how critical mathematics writing unfolds from the perspectives of the young students and those of the pre-service teacher students (Vivaldi, 2014). This discovery, by its very essence, is what taps a great deal of profound, higher-order thinking of pre-service teachers and students.

Once again, the problem is as follows: "Farmer Jones has chickens and cows on her farm. When she looks under the fence, she counts 8 heads. When she looks under the fence, she counts 22 legs. How many chickens does Farmer Jones have? How many cows does Farmer Jones have?" In the process of investigating the different approaches young students took to solve the mathematical problem presented, Angela asked one eighth-grade student and one student who was a junior in high school. She informs us from the outset that she knew exactly what the eighth-grade student was thinking when attempting to solve the chicken-and-cow problem.

> **Eighth-grader's writing:** ... considering the farmer had 8 animals, I took all the numbers that added to 8 and made a chart with 2 columns, cows and chickens. I multiplied the cows column by 4, since cows have 4 legs. I multiplied the chickens column by 2, because chickens have 2 legs. I then added the totals across and figured out which combination would give me a total of 22 legs.

Angela writes

> **Angela's writing:** I noticed when the eighth-grade student solved this problem, they used the same method that I did when I completed this problem in class. They took the information given with the farmer seeing 8 heads and 22 legs and made a table of how many legs each animal would have according to the numbers given.

Not only does Angela identify the similarities of procedures in solving this problem; she also indicates the method that was used—putting data in a table. She goes on.

> **Angela's writing:** They used the method of guessing and checking and making a table to test out their answers which led to the correct answers of how many of each animal there were. (They are listed as student number one in the photo provided below). By using this method to solve the problem, they came up with it being 3 cows and 5 chickens (See Figure 7.4).

It is safe to say, then, that Angela found this student to be using a guess-and-check, or trial-and-error, approach to solving the chickens-and-cows problem. More specifically, the student started with 0 cows and 8 chickens to arrive at 16 legs. But there were 22 legs altogether, so, they then went on to 1 cow and 7 chickens to get 18 legs. Since the answer was becoming closer and closer to the correct answer, they continued with 2 cows and 6 chickens. This yielded 20 legs. Thinking they were almost there, they then tried 3 cows and 5 chickens, which finally produced the right answer of 22 legs. Angela then explains what she found when she observed the student who was a junior in high school attempt to solve the chicken-and-cow problem.

> **Angela writing:** Interesting enough, the high schooler took a different approach to solving this problem when presented to her. The high schooler used the concept of variables and created a system of equations to solve this problem in an algebraic manner. She used x to represent the number of chickens on the farm and y to represent the number of cows there are. She then created two systems of equations: $x+y=8$ and $2x+4y=22$. She then solved each equation and got her answer. (She is listed as student number two in the photos provided below).

So, as shown in Figure 7.5 below, the high school junior wrote the following in her explanation of her thought process in attempting to solve this problem. Again, it is necessary to repeat the multilayered nature of asking a pre-service teacher student to observe K-12 students solve mathematics problems because doing so will allow the reader to experience both the younger student's written explanation and the pre-service teacher's written interpretation. The high school junior writes:

> **High school junior's writing:** I created a system of equations knowing the information given. I labeled chickens, x, and cows, y. In the first equation I wrote was $x+y=8$ and because it was stated that there were 8 animals in total. The second equation was $2x+4y=22$ because chickens have 2 legs and cows have 4. I made $x+y=8$ into $y=8-x$ by subtracting x on both sides. I then plugged the $y=8-x$ into $2x+4y=22$. I solved the equation and found that $x=5$. Using the x I plugged it into $x+y=8$ and solved for y and found $y=3$. I checked my work by plugging the x and y values into the equation (See Figure 7.5).

What is even more inspiring with regard to Angela's math writing is the way in which she explains the implications of what she observed.

> **Angela's writing:** After completing this investigation, I noticed that as a student moves up grade levels and gains more and more experience in solving mathematical equations, they complete problems much faster and much more efficiently in terms of algebra. Each student had their own unique way of solving it and both came up with the right answer. This investigation will help me in the future concerning my teaching career because it taught me that if a student is comfortable using a certain method when solving a problem, I should give them the flexibility to do so at least for a little while before they get familiar with another method that they just learned. By doing this, I can make the already stressed student a little less stressed by allowing them to use a method they are comfortable with and that they know how to do well. I want to try and be both an accommodating and effective teacher by easing students into new content instead of just throwing it at them.

Mathematics writing can also involve pictures. For example, my other students worked with first graders who solved the chickens-and-cows problem by drawing circles for heads and line segments for legs. These young students initially place two line segments below each circle because they know that animals have an even number of legs (i.e., 2, 4, 6, etc.). Thus, they arrive at 16 legs. To account for the cows, they simply add an additional two legs on three circles and arrive at 22 legs (See Figure 7.6).

A Note on Mathematics Standards

It should be noted that, as in writing critically in the natural sciences, an important premise when writing critically in mathematics is to think and write critically about mathematics standards. Whatever one's position is about federal education standards, it is important to reconsider carte blanche acceptance of them. In the next chapter, I elaborate on how federal standards in education are tools of submission and can be used for purposes of alienating children and older students and even teachers.

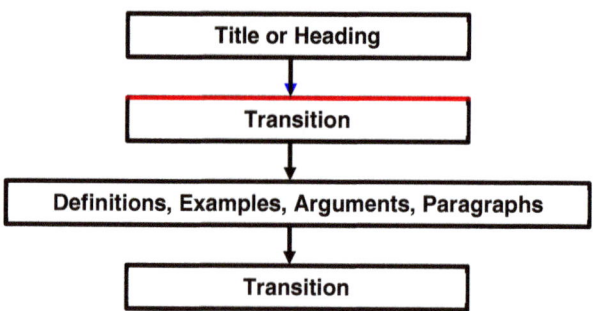

Figure 7.1. Layout and Structure of a Mathematical Segment

WRITING CRITICALLY IN MATHEMATICS

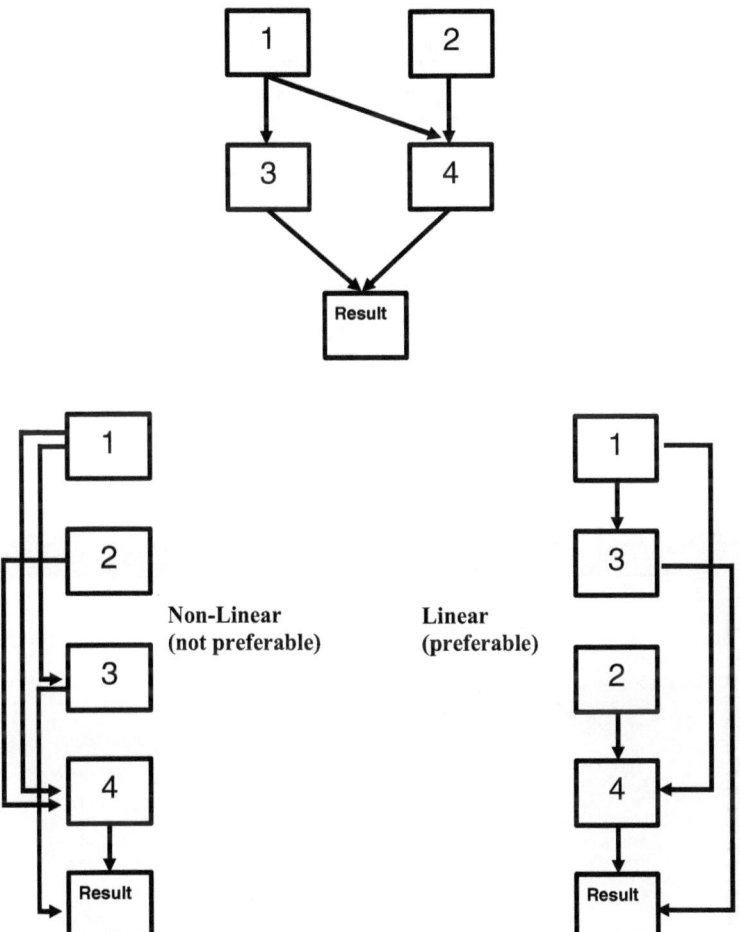

Figure 7.2. Dependency Graph and Flow of Specific Arguments

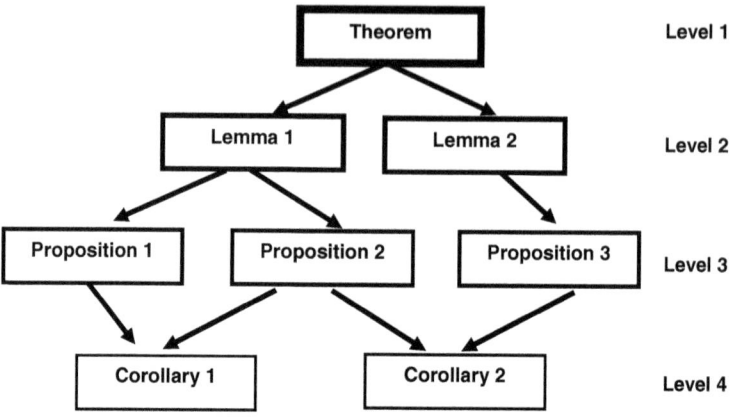

Figure 7.3. Mathematical Hierarchy of Form

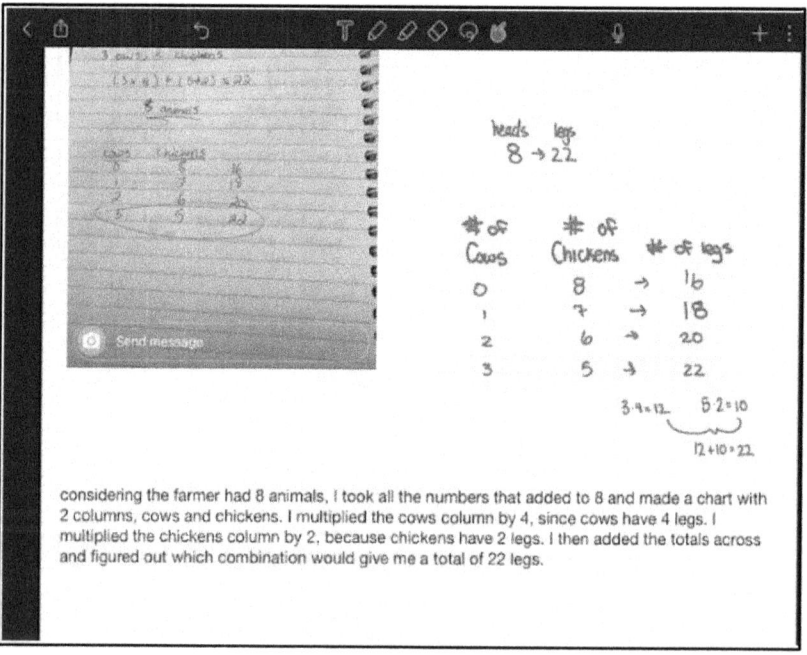

Figure 7.4. Eighth Grader's Analysis of the Farmer Jones Problem

WRITING CRITICALLY IN MATHEMATICS 103

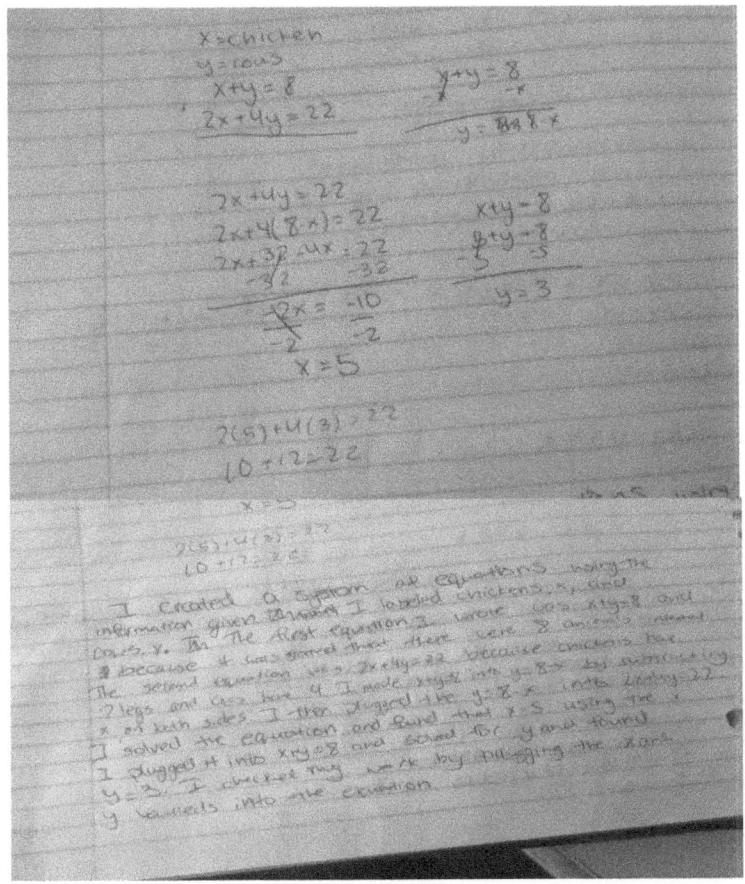

Figure 7.5. High School Junior's Analysis of the Farmer Jones Problem

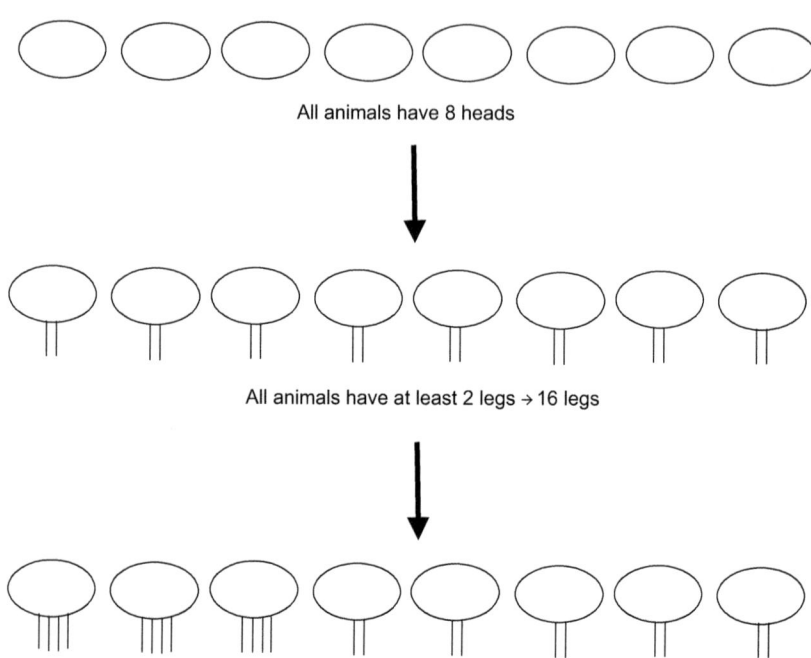

Figure 7.6. First Graders' Analysis of the Farmer Jones Problem

· 8 ·

QUESTIONING NORMS IN CRITICAL STEAM WRITING

At this point, we have become familiarized with the different ways in which one can approach writing in STEAM critically by focusing each of the last four chapters on writing approaches in each of the STEM disciplines. In reading these chapters, you may have noticed a pattern. And that is, when STEAM students and professionals write, they do so across STEAM subject areas, and this occurs regardless of intention. The natural scientist is more often than not required to introduce a mathematical explanation for something that occurs during an experiment or possibly when writing a theoretical paper. The mathematician will often have to address concepts in the arts, including architecture (Lange & Lange, 2012)—especially when it has to do with dimension and measurement—and philosophy or history. The engineer might need to consult an article in a physics journal—especially today, in the current cyber-intensive zeitgeist, when semiconductor-related technologies are increasing possibilities for human existence. For example, buildings are increasing in height and size, many of which are in earthquake zones, thus making it possible to increase habitability in traditionally hazard-prone or precarious locations. The main point here is that it is relatively difficult not to write with related STEAM subject areas in mind.

In this chapter, we continue our examination of critical writing in STEAM disciplines by addressing key issues that confront students and their teachers in STEAM throughout the educational system. Note that these issues exist in non-STEAM areas as well, so it will be important to engage them explicitly in a way that demonstrates criticality and impartiality. Initially, we will outline the need to write in a way that addresses each reader, regardless of that reader's multiple identities. Being critical and impartial necessitates all STEAM writers to accept and adhere to antiracism, anti-sexism, and the many additional "anti-isms" that respect all individuals and identities in our postcolonial society. And second, we will place STEAM writing within the context of educational settings and ask ourselves if federal and state interventions—i.e., the standards—are the right direction to take if we are to ensure every child's and adolescent's academic transcendence. And finally, we will examine how writing in STEAM can adversely affect student outcomes when it comes to educational assessment or evaluation. Our plan, then, will be to examine alternative models to assessment that prizes each student's personal histories and experiences as a direction toward student progress.

Equity in STEAM and Writing

Up to this point, we stressed the importance of avoiding masculinization of writing across the board, not solely in STEAM disciplines. In Chapter 2, stress was placed on the importance of writing using an antiracist and gender-fluid approach—one that underscores not only the avoidance of antiquated terminologies but also the need to listen to voices of those who have been traditionally excluded from so-called mainstream society. In Chapter 6, we showed the importance of rejecting gender-based assumptions in the context of email and letter writing. Mistakes of this kind can fall anywhere between (and including) self-consciousness and outright humiliation.

From an international position, the United Nations Educational, Scientific and Cultural Organization (UNESCO) has chosen to foster educational policies that focus on inclusion, literacy, quality, capacity development, and finance. Long-term international, national, and local commitments are needed to help increase participation, equity, and quality in education, particularly in the most poverty-stricken and marginalized communities. UNESCO has also continued to promote gender equity and gender equality in education (2019). In 2005, gender parity at both the primary and secondary levels occurred in only 59 out of 181 participating countries worldwide. Within these remaining 126 countries, 63% have achieved gender parity at the primary level, 37% at the secondary level, and 3% at the tertiary level. These gender disparities are more prevalent among those in regions with the highest poverty rates. Once educational access is achieved, girls tend to stay in school longer and either compete with or outperform boys, particularly at the secondary level. However, severe gender biases still affect girls in places such as Afghanistan, Yemen, and in most nations of sub-Saharan Africa. At the tertiary level, for the first time, more women than men are entering higher education institutions in many parts of the world. However, women are still underrepresented in actually or perceptually male dominated fields such as engineering, natural sciences, mathematics, and agriculture—STEAM fields for the most part. Equity in education will require greater effort to address race and gender disparities throughout the world so that each and every student has a voice (Ayers, 1993; Blake, 1995). Local, national, and international efforts must continue to identify and resolve race and gender-biased practices that have traditionally compromised efforts of advancing equity.

Is Adherence to National Standards and Writing Critically in STEAM Contradictory?

Nearly everyone would agree that in most fields and occupations, standards are necessary to hold experts and professionals accountable for their practice or research. Indeed, we would expect that our doctors meet a baseline set of criteria; they should categorically swear by the Hippocratic Oath to *do no harm*. And we certainly don't want our doctors to engage in iatrogenesis—the deliberate or unwitting causation of a malady, a harmful complication, or other adverse effect by any medical activity ranging from diagnosis to intervention, error, or negligence. In a similar manner, we would want our engineers to follow guidelines that are based on the universal laws of physics in three-dimensional space for application in the real world; they would need to know how to take the necessary steps to prevent structural failure when strong winds or other calamitous natural events occur. We would also want our commercial airline pilots to be experts in engineering planes so that they take off, fly, and land safely without any equivocation or reason for pause. Expecting these outcomes from experts on whom we rely in our everyday life in situations on which we depend in life-or-death situations is not thinking in the extreme. In short, these professions depend on the expert to follow tried-and-true procedures for ensuring the safety of anyone who needs medical care, walks into any building, or travels on a plane.

But, when it comes to national standards, there are other acutely important professions in society that *cannot* be likened to medical experts, structural engineers, and airplane pilots; universal guidelines would only serve to impede progress and not facilitate or advance it. One such profession is the educator. Indeed, many careers fall into this category, some of which include the schoolteacher, college professor, early childhood teacher, vocational instructor, and mentor. If we take a step back from the ideas that standards have good intentions, it might be important to investigate whether standards, when used in the extreme, can either negatively influence or outright harm student success and progress.

But, like many things taken to their extreme, nationalized educational standards can be powerful tools for exploitation (Farenga & Ness, 2017). Upon reflecting on nearly two decades of research examining education policy, I am concerned that adverse influences of market-based education reform in addition to state mandating of teacher accreditation have waxed and not waned. While the academy has been relatively passive and dormant in terms of its stakes in teacher preparation, the corporate world has been stealthily active in its quest to make public education private. One of the current phenomena in the United States system of education that has stifled and imperiled the welfare of students is the market-based national standards movement through the medium of Common Core State Standards (CCSS). And ever since the late 1980s, when national standards were slowly emerging as a national phenomenon, it appears that punitive measures against educators through standardization in education has increased considerably—to the point that it has begun to adversely affect student outcomes and overall academic progress.

Paulo Freire (1970) and Ivan Illich (1972) have provided their own explanations and perspectives on issues related to credentialing and certification, issues that parallel the universality of standards in our own time. Illich challenged the belief that credentialing and certification were necessary. For Illich, schooling devoted to achieving consensus-driven benchmarks through credentialing and certification has perpetuated a sense of hopelessness for underprivileged youth. Rather than preparing students for a world in which active engagement, motivation, curiosity, and collaboration has become the paradigm of academic and social advancement in the 21st century, dependence and the perverse overuse of standards has led to students with fixed mindsets who lack motivation in school, and who develop the belief that education has little, if any, use in their daily lives.

The standards serve as the structures to support domination over programs, disciplines, faculty, and students. After critical analysis of the standards, questions of academic freedom and liberation fill the void of critical consciousness caused by the uncritical carte blanche acceptance of standards. The standards and their prescriptions have contributed to what Freire (1970/1996) called "the culture of silence" (p. 12), which is characterized by economic, social, and political domination, and is supported by the false notion that the reason for poor achievement is predicated on the development of poorly trained teachers. Since the publication of A Nation at Risk in 1983, the public has putatively accepted this perception. Those whose interests are served by this myth have ignored the escalating culture of poverty that exists in this country and instead have persistently intimidated the academic community over reports of a population's poor performance. The stentorian, yet flawed and unsubstantiated, message of the intersection of the neo-conservative and neo-liberal education agenda is consistent and clear: Poorly trained teachers cause students' poor academic achievement.

Critics who are in favor of standards often ask me to provide data that demonstrate evidence supporting national standards as obstacles to student progress. My response in this regard has been the same year after year—simply examine the results in international assessments, such as PISA or TIMSS and explain to me why the United States always fares mediocre or poor in reading, writing, working with numbers, and thinking scientifically. But the evidence goes well beyond just numerical data; rates of cognitive growth in learning any academic or vocational subject cannot be measured using norm-referenced procedures—that is, assessments, which only measure success through comparison. The best way to consider student (or teacher) cognition is through one-on-one cognitive interviews or criterion-referenced assessments that assess the extent to which the individual understands a concept by completing a cognitive task. What is worse is that nationalized standards have been increasingly focusing attention on dispositions and, unfortunately, the efforts of developers of standards have failed with regard to the creation of rubrics that are intended to measure disposition.

The contemporary standards-based assessment movement relies on rubrics—scoring guides in which their designers attempt to operationalize a set of standards or objectives for application in evaluating students' work. Our focus in the remainder of this final chapter is to examine the role of rubrics as they are applied to writing in STEAM, and to learn how they are ill-equipped to measure the strengths and weaknesses of STEAM papers. We close by considering equitable ways of assessing the strengths and limitations of STEAM papers.

Rubrics are destructive for STEAM writing students for several reasons. Criticisms of rubrics in the education research literature are ubiquitous (Andrade, 2005; Delandshere & Petrosky, 1998, 1999, 2002; Hillocks, 1997; Johnson, Johnson, Farenga, & Ness, 2005; Farenga, Ness, & Sawyer, 2015; Koretz, 2009; Mabry, 1999; Moskal & Leydens, 2000). To begin with, rubrics either underestimate or overestimate writing ability, which provides students or their parents with an inaccurate or distorted evaluation. Thus, rubrics often fail to reflect a student's skill level. Second, rubrics are often used to quantify either complex behaviors or dispositions—characteristics of academic subjects that simply cannot be evaluated by a single number or term (Delandshere & Petrosky, 2002; Mabry, 1999). Koretz (2009) argues that separating students into categories, such as "below basic," "basic," "proficient," and "advanced," is a potentially misleading measurement and is "one of the worst decisions we made in testing in decades" (p. 2). I suggest that the current assessment practices that are supported by state departments of education avoid transparency by using rubrics in order to obfuscate the evaluation process. Third, much of the language that is used in a rubric's design often lacks clarity of definition. This vagueness in wording is a result of the futility to operationalize specific terms and phrases. In other words, in professions that might require specific guidelines for properly and effectively conducting procedures, rubrics must be defined in a way that directly corresponds with successful completion of a task. For example, within the health professions, rubrics have been used for assessing research skills, assessing one's ability to present on a scientifically demanding topic, improving the quality of online instruction, participating in online discussions, determining clinical performance in an operating room or emergency room, and assessing skill development of aseptic practices. Taken from Brown, Conway, and Sorenson (2006), Table 8.1 shows a partial sample (three out of five components) of an operationalized rubric for students in the pharmaceutical professions.

Table 8.1. Aseptic Technique Scoring Rubric

Component	Ratings		
	Likely Harmful	**Needs Improvement**	**Acceptable**
Personal preparation	Jewelry not removed OR hands washed for less than 30 seconds OR germicidal soap is not used OR faucet is turned off without using toweling OR gloves not donned	N/A	Jewelry removed AND hands washed for at least 30 seconds with germicidal soap AND handles of faucet turned off with toweling AND gloves donned
Disinfecting hood	Hood not wiped with isopropyl alcohol and sterile 4x4s OR hood wiped in a manner that does not move from cleanest area to dirtiest area OR some area of hood missed during process	N/A	Hood wiped with isopropyl alcohol and sterile 4x4s from cleanest to dirtiest area without missing any area
Working in airflow	Airflow from HEPA filter across access points or needle is routinely blocked (student blocks airflow and does not quickly correct) OR manipulations occur within first 6 inches of hood	Airflow from HEPA filter across access points or needle is rarely but occasionally blocked (student blocks airflow but quickly corrects)	Airflow from HEPA filter across access points and needle is never blocked

Notice that the designers of this rubric use extremely precise language. In fact, the rubric is so unyielding that two out of the three components included have only two, not three, ratings—"Likely Harmful" and "Acceptable." There is no middle ground when evaluating one's personal preparation and ability to disinfect something. Also, there are fewer ratings in rubrics designed for professions that need to follow strict guidelines for correct practice. The terms in this rubric are operationalized. For example, "hands washed for less than 30 seconds" is clear, concise, and definitive; that is, the designer does not state "hands are partially washed" or "uses some water." Clearly, the words "partially" and "some" would need to be operationalized with specificity. The idea here, then, is that the rubric works for those who are studying to practice pharmacy.

The use of rubrics in this instance, however, ends with their inclusion in specific disciplines, many in STEM, that require such an approach for purposes of safety and wellbeing. Rubrics for writing in general and writing in STEAM in specific is another issue altogether. Like the use of rubrics in art and music, rubric use in writing is extremely harmful as it leads to decreased motivation, normed hierarchies in which students are categorized as "able" and "incompetent," and student acquisition of the fixed mindset disposition—one that places greater emphasis on epistemological innateness (e.g., "I'm not a math or science or writing person . . .") than on effort and hard work.

Lastly, rubric assessment is pernicious because it supports oppression, thus limiting the future freedom of opportunity for the STEAM student and professional writer (Ho, 2008). Further, many rubrics that are designed with the intention to enhance the assessment process do precisely the opposite—they supply little, if any, additional data for instructors and students. In reality, most rubrics are not similar to the example in Table 8.1. Quite the opposite, they are designed as semantic puzzles whereby the evaluator is obligated to determine how to apply ratings that have different shades of meaning. Table 8.2 demonstrates how rubrics are often used in education, particularly in writing as it is applied to disciplines in STEM, social sciences, and the humanities. This writing rubric was designed by the Northwest Regional Educational Laboratory (NWREL) as purportedly a way to assess six traits of writing that the designers believe students should possess. The entire rubric is composed of several key questions. The key question in this example has to do with written expression: Does the organizational structure [of the manuscript] enhance the ideas and make them easier to understand?

Table 8.2. NWREL Writing Rubric—Organization

Component	Not Proficient			Proficient		
	1 Beginning	2 Emerging	3 Developing	4 Capable	5 Experienced	6 Exceptional
General	No identifiable organization; writing lacks a sense of direction or seems random	Uses mostly ineffective organization with only a few sections that direct reader	Shows developing organization that is sporadic, hampering ability to follow text	Includes basic organization that moves the reader through the text logically with minimal confusion	Reflects smooth, cohesive organization and varied techniques that build to create a unified whole	Develops seamless organization that enhances and showcases central ideas and their relationships
Lead and Conclusion	Uses no lead and no conclusion	Includes either a lead or a conclusion. If lead is present, fails to establish purpose; if conclusion is present, fails to provide closure	Includes both a lead and a conclusion. Lead fails to establish purpose, and/or conclusion fails to provide closure	Contains a lead that establishes purpose, though may be formulaic or obvious. Contains a conclusion that provides closure, though may be formulaic or obvious	Features a lead that creates anticipation and establishes clear purpose. Includes a conclusion that ties up loose ends, providing a satisfying sense of closure	Creates an inviting lead that establishes clear purpose, draws a reader in, and creates a strong sense of anticipation. Develops a satisfying conclusion that conveys a powerful and thoughtful sense of closure
Transitions	Lacks use of transitions Does not use paragraphs	Rarely uses transitions Uses paragraph breaks sporadically	Uses transitions that are repetitive, inconsistent, and/or fail to connect ideas Separates ideas into paragraphs weakly	Includes transitions that connect ideas, though they may be formulaic or predictable Consistently separates ideas into distinct paragraphs	Features logical, varied transitions that connect and develop ideas Includes paragraphing that supports ideas	Features thoughtful, smooth, varied transitions that clearly connect ideas and enhance meaning Uses paragraphing that enhances ideas

Continued

Table 8.2. Continued

Component	Not Proficient			Proficient		
	1 Beginning	2 Emerging	3 Developing	4 Capable	5 Experienced	6 Exceptional
Sequencing	Uses no sequencing of ideas	Uses very limited sequencing that fails to show how ideas fit together	Uses sequencing that fails to showcase ideas or becomes formulaic	Provides logical sequencing of ideas	Employs sequencing that builds connections to create a unified whole	Utilizes highly effective sequencing, making best choices for progression and enrichment of reader's understanding
Pacing	Uses no evident pacing	Uses uneven pacing, slowing down or speeding up inappropriately or awkwardly	Appropriately controls pacing in some sections but not in others	Evenly controls pacing in most places	Exhibits well-controlled pacing throughout	Skillfully uses pacing to compel the reader through the text and enhance its impact
Purpose/text structure	Uses no discernible text structure or purpose	Uses loose text structure that leaves reader unclear or confused about purpose	Uses text structure inconsistently, affecting the reader's ability to identify purpose	Uses text structure consistently to reflect purpose, moving the reader through the text logically, with minimal confusion	Employs text structure that clarifies and supports purpose throughout	Utilizes text structure that enhances understanding of purpose and flows very smoothly
Title	Uses no title	Uses a title that does not link to main idea or is misleading	Uses a title that is formulaic, nondescript, or fails to link directly to main idea	Includes a title that connects adequately to main idea	Selects a title that reflects main idea in an unusual or interesting way	Draws the reader in with an original title that reflects main idea and captures deeper meaning

The NWREL rubric suffers from a number of factors, which can adversely affect student outcomes. First, nearly the entire rubric contains words and phrases that appear to be mentalistic. In other words, each of the rubric's items is sensitive to individual impressions and attitudes; few, if any, terms can be operationalized, and therefore cannot be measurable. For example, a writer who has a capable grasp on organization, according to this rubric, will "... include basic organization that moves the reader through the text logically with minimal confusion ..." Indeed, this item is packed with non-measurable language. What is "basic organization?" How can one know for certain that "basic organization" (however this is defined) "moves the reader?" Moreover, what does it mean to "move" a reader? In addition, what in specific makes text logical enough to move a reader? And what is the threshold of "minimal confusion?" Second, the very idea that the items in this rubric are sensitive in meaning to the individual user is a telling signal that it is impossible to assess writing with one-size-fits-all models like rubrics. That is, it is impossible for one's writing to "move the reader" in a quantifiable or measurable way, but it *is* possible for one's writing to move the reader. Third, there are minimal, if non-existent, differences between and sometimes among the ratings of this rubric and nearly all rubrics like it that are designed for the purpose of measuring one's writing propensity. Shades of difference are sometimes indiscernible altogether. For example, a writing student who receives a rating of "capable" for sequencing paragraphs is said to "... provide logical sequencing of ideas ..." Another student who receives a rating of "experienced" (which, according to this rubric, is at a higher level than "capable"), is one who "Employs sequencing that builds connections to create a unified whole ..." It is, therefore, hard to know whether "logical sequencing of ideas" is better or worse than "building connections to create a unifying whole."

And lastly, words such as "proficient," "beginning," "emergent," "developing," "experienced," and the like—all terms with different meanings—have all been used to describe STEAM writing students who are frequently led to believe that their current level of effort and persistence will be sufficient when engaging in future activities. However, evidence suggests that this is not the case. One need only examine the number of higher education institutions that have a proliferation of remedial courses in writing and literacy overall. Many students often graduate with "A" or "B" averages and have passed all of their state assessments. Paradoxically, a majority of incoming freshmen are unprepared as writers and, moreover, their academic gains are limited as their years progress. Arum and Roksa (2011) conclude that after three semesters of college, students barely show noticeable gains in the areas of critical and complex reasoning and written expression.

The point? Critical writing in STEAM, or any other discipline for that matter, presupposes that one's aptitude in writing is not based on a number or any other quantifiable result. Just as a composer's piece of music or an artist's painting cannot be judged by numbers, so too the STEAM writer cannot and should not be overwhelmed by specific universal standards that may or may not overlap with the narrative that asks: What is the best way to help our writers (of STEAM or not) improve their craft?

Alternatives to Standards and Rubrics

In brief, we conclude this text by identifying some important alternatives to nationalized standards and rubrics that will help teachers and students evaluate writing, particularly writing in STEAM. These alternatives include the following: ratiocination by writing genre; writing by emulating models; peer evaluation; and concepts and methods that lead to self-understanding and self-fulfillment.

Ratiocination by writing genre is a useful method of determining the virtues and obstacles of STEAM students' or professionals' written work. Indeed, different genres of writing possess different structures and forms. The engineer's written letter possesses a different style when compared to the engineer's business proposal or research paper. The same holds true for the variety of styles of written work produced by natural scientists, mathematicians, architects, computer scientists, and professional technologists. So, for the STEAM researcher, it will be essential to understand the general outline of the research paper that was included in Chapter 4, namely, the overall structure of "Introduction," "Method," "Results," and "Discussion," as well as each of their subheadings. This knowledge will enable the instructor to ratiocinate the paper more easily because the paper's structure through the use of appropriate headings will help guide student writers in developing their unique narratives, findings, and perspectives. Instructors should support their STEAM students by providing the correct organization and styles of various written genres that students will undoubtedly be exposed to.

Writing by emulating models is a second way to take into account STEAM writers' works. Before identifying what "writing by emulating models" is, it might be useful to initially identify what it isn't. Writing by emulating models is *neither* copying someone else's writing verbatim *nor* is it paraphrasing someone else's writing without acknowledging that author's work. Let's be very clear: The act of copying someone's work verbatim is plagiarism and penalties for engaging in this action range from course failure to institutional expulsion. Quite the contrary, one who successfully emulates models examines an instructor's or favorite author's or researcher's written work and identifies its strengths in form, structure, and delivery. By emulating some of your favorite authors, you will learn how to avoid your own idiosyncratic writing. Examples of idiosyncratic writing abound. I know from my own experience that in my undergraduate college years, I would often have the habit of writing words and expressions repetitively, throughout my papers. Some of the expressions that I remember writing more than once, even in the same sentence and the same paragraph, were "... and so forth," too many "in additions," and "as such" or "accordingly." Student and even professional STEAM writers will often engage in idiosyncratic writing by repeating grammatical and syntactical errors, specific words, or use all too common idioms or clichés. All too often, I read students' papers that are filled with run-on sentences or the opposite—sentence fragments. I also see many students, even those who are skillful in writing, employ misplaced modifiers: "After mixing hydrochloric acid with sodium hypochlorite, the result was a violent chemical reaction that produced a great deal of heat and gas." Indeed, the "result" did not do the "mixing." Students who emulate their favorite writers will notice that both their content and their sentence and paragraph structure is consistently correct.

Third, peer analysis serves as an excellent model for written discussion—especially for STEAM students and professionals. When my students present their simulated lessons in front of class, I ask the students who are not presenting to provide constructive commentary that will help the student presenter critically reflect on their performance and identify their own desirable qualities and shortcomings. In corroboration, Chase, et al. (2020) posit "that there needs to be reflection-in-writing wherein teachers and teacher candidates reflect in the moment they are writing, as well as reflection-on-writing, which involves peer- and self-review of written work and opportunities to revise" and further argue that "... reflecting upon what is written, why it is written, and how one feels while composing involves thinking beyond oneself to account for context" (p. 66). This format also works well when examining peers' written work. A well-known adage relating to expertise is that one who teaches learns more when teaching than when passively listening because when you teach, you consolidate and reinforce your knowledge even more than you do when learning without active engagement. Peer review analysis can be done in pairs, in groups of three or more individuals, and as whole assessment.

Small-group analysis can range in size of three to six or seven students. Several options are available for small-group peer analyses. These options include, but are not limited to, having students to read and respond to specific writing passages or genres in class, and having students read written passages or different types of STEAM genres posted on course websites on which they will need to prepare comments in advance. The number of options an instructor provides for group peer assessment will be dependent on the type of structure and size of the small groups. Students should not be expected to evaluate passages over three or four pages for each class section as this will take up too much class time and possibly overwhelm the cognitive load of each student.

Full-class analysis of writing is useful for developing shared ideals about specific attributes on which to focus when reviewing a paper as well as what tone to use when providing constructive feedback. This structure of constructive feedback for the purpose of analysis is helpful because it provides students with a wealth of ideas that will help them revise their work using a variety of concrete examples from essays under development. This method of peer analysis is most successful when students have prepared feedback on the paper before class and are prepared to examine the paper in detail. With this type of examination, instructors should prepare their own feedback on the paper and formulate a plan of action for leading discussion. In order to ensure that students have more exposure to their peers' work, instructors can schedule more full-class peer analyses.

As with any approach to peer analysis, it can be helpful to ensure that students receive constructive feedback from more than one peer on any given written assignment. Doing so provides them with a better idea of whether a particular reader's perceptions of their work are likely to resonate with other readers. Instructors who prefer small groups throughout the term will need to decide whether students should engage in peer analysis in the same small groups regularly, or whether it would be more desirable to alter the membership of the groups from one written assignment to the next. Students who are kept in the same groups throughout a term tend to develop a sense of anticipation in terms of expectations of constructive comments and feedback. At the same time, it might also be useful to allow students to work with different writers so they can receive a greater variety of comments and possibly a greater likelihood of impartial feedback.

Lastly, I provide probably the most important consideration, namely, an alternative to the adverse consequences of assessment and evaluation. To be sure, assessment and evaluation has impaired education to the extent that students avoid genuine learning and conceptual understanding due to the deleterious effects of student comparisons at both the micro (student-to-student or school-to-school) and macro (country-to-country—cross-cultural comparison) levels. As a society, we can abolish achievement gaps and provide pathways to eradicate white supremacy, cisnormative heteronormative patriarchy, ableism, imperialism, colonialism, systemic racism, sexism, xenophobia, ageism, and homophobia by dispensing with assessment and evaluation altogether. So, rather than self-assessment or self-evaluation, a better alternative for our STEAM and non-STEAM students and professionals would be to focus instead on self-understanding and self-fulfillment. To this end, STEAM writers—and non-STEAM writers as well—can use the concept and method of *currere* as a key to self-understanding.

I should point out that providing a synopsis of *currere* would do a disservice to the reader given that the study of *currere* and its practice is something that takes months and even years to accomplish successfully. That being said, I provide an explanation of *currere* so that readers can expand on their knowledge and utilization of *currere* by referring to other important sources on the topic. It is important to note that this explanation in no way conveys the totality of *currere* and by no means should be used in such a way.

Initially developed by Pinar (1975, 1995, 2004), the concept and method of *currere* can serve as an autobiographical and autoethnographic approach toward a more enlightened way of thinking in any area of inquiry (Wang, 2017, 2020). *Currere*, which, in Latin means "run the course," provides the individual with an action verb alternative to the word curriculum, a noun, which means "course" in Latin. Given that no two people learn the same way, the individual, then, actively engages in self-understanding through the examination and analyses of one's own experiences. The practice of *currere* consists of four chronological steps: the regressive; the progressive; the analytical; and the synthetic. In the regressive, the individual examines oneself by elaborating personal prior experiences that serve as a guide for contemplation and reflection of how past events or occurrences may have shaped one's attitude and knowledge and had effects on one's learning. Upon realization of past experiences, the individual can then determine how to direct oneself in future situations and events. To this end, the progressive step hones the possibilities of these situations and events. The progressive leads not only to what the future can be, but how to arrive there and what it might look like. While Pinar urges that these steps are not linear, placing the progressive second in the sequence allows for more expansive thinking and a greater consciousness and role of the progressive when considering the future. In the analytical, the individual examines the here and now as a means of creating a subjective space to transcend the present. The final step, the synthetic, allows the individual to engage in personal transformation—to revisit positive experiences and dispense with negative ones for the purpose of developing new spaces. Wang (2022) informs us that teachers can use *currere* to examine personal experiences in the classroom by tapping their students' psychic speech, which includes each student's ways of knowing through their own symbolic representations.

While it might take time for the teacher to practice it correctly, *currere* is a concept and method that teachers and students should eventually use as a means of transcending any area of inquiry for the purpose of self-understanding. In the following list, I have provided sources that expound on *currere* so that writers can develop their abilities in using currere for the purpose of self-understanding, self-examination, and self-fulfillment. Note that this is by no means an exhaustive list, but one that should get writers started on their way to practice *currere*.

Introductory List of Sources on Currere

> Baszile, D. (2017b). On the virtues of currere. *The Currere Exchange Journal, 1*(1), vi–ix.
> Kimiecik, J. (2022). The feel of currere. *The Currere Exchange Journal, 6*(1), 63–73.
> Kincheloe, J. (1998). Pinar's currere and identity in hyperreality: Grounding the post-formal notion of intrapersonal intelligence. In W. Pinar (Ed.), *Curriculum towards new identities* (pp. 129–142). Garland Press.
> Pinar, W. (1975). Curerre: Toward reconceptualization. In W. Pinar (Ed.), *Curriculum theorizing: The reconceptualists* (pp. 396–414). McCutchan.
> Pinar, W. (2004). *What is curriculum theory?* (1st ed.). Lawrence Erlbaum Associates.
> Pinar, W. (2019). Currere. In J. Wearing, M. Ingersoll, C. DeLuca, B. Bolden, H. Ogden, & T. M. Christou (Eds.), *Key concepts in curriculum studies: Perspectives on the fundamentals* (pp. 50–52). Routledge.

Sawyer, R. D. (2022). Re/membering curricular entanglements: A currere of the present-absent curriculum of a gay high school student. *Journal of Curriculum Theorizing, 37*(1), 23–38.

Smith, B. A. (2013). Currere and critical pedagogy: Thinking critically about self-reflective methods. *TCI (Transnational Curriculum Inquiry), 10*(2), 3–16.

Wang, W. (2017). Currere, subjective reconstruction and autobiographical theory. *Transnational Curriculum Inquiry, 14*(1–2), 110–141. https://doi.org/10.14288/tci.v14i1-2.188678

Wang, W. (2020). *Chinese currere, subjective reconstruction, and attunement: When calls my heart*. Palgrave Macmillan.

Wang, W. (2022). Currere, psychic speech and teacher education. *The Currere Exchange Journal, 6*(2), 18–26.

Conclusion

I would like to close this chapter, and this book, by referencing back to something I had pointed out at the outset—something that is accurate but probably is not entirely contextual or circumstantial. You may have inferred from the end of the Introduction that since we started on a voyage at the beginning of the book, we would steer the ship back to the mooring at the port by book's end. This inference is not entirely correct: You joined me on this voyage of STEAM writing at the beginning of the book, but the voyage does not end here. Rather, our investigation in STEAM writing should be the commencement of a lifelong voyage, full of writing adventures that are convincing, persuasive, motivating, and exciting for your readers. So, I now hand the torch, or pass the mantle, over to you so that you can empower your STEAM students, colleagues, or clients with the skill of critical STEAM writing.

GLOSSARY OF TECHNICAL TERMS IN WRITING AND STEAM

ACS. American Chemical citation style; Style Guide published by the American Chemical Society. ACS is the format commonly used in the field of chemistry and some of the tangentially related natural sciences within biology, geology, and physics. ACS is a numbered style with references numbered in the order of appearance in the article and listed in that order at the end of the article.

abstract. A succinct, independent summary of the contents of a research paper that appears independently at the beginning of the paper. Abstracts may also appear as summaries of the contents of a paper presentation or poster.

acknowledgments. A section preceding the references (or a footnote, or a final paragraph) in which the author thanks people or organizations for help, advice, or financial support for the work described.

ACM. The Association for Computing Machinery (New York, USA).

AIP. Style refers to the citation format established by the American Institute of Physics. AIP is the format commonly used in the field of physics. AIP is a numbered style with references numbered in the order of appearance in the article and listed in that order at the end of the article.

Algorithm. A procedure or set of rules that must be followed in order to carry out an instruction for fulfilling a command. Algorithms are common in mathematics, the natural sciences, and in computer coding.

APC. Article processing charge. A fee paid to publish an article in an open access journal or hybrid journal.

AMS. American Mathematical Society Style Guide

APA. American Psychological Association Style Guide

ASCII. American Standard Code for Information Interchange. A coding system in which letters, digits, punctuation symbols, and control characters are represented in seven bits by a number from 0 to 127. An eighth bit is often added to allow extra characters.

bibliography. A list of publications on a particular topic, or the reference list of a book.

BibTEX. A program that cooperates with LATEX in the preparation of reference lists. It makes use of bib files, which are databases of references in BibTEX format.

citation. A reference in the text to a publication or other source, usually one that is listed in the references.

conjecture. A statement that the author believes to be true but for which a proof or disproof has not been found.

copy editor. A person who prepares a manuscript for typesetting by checking and correcting grammar, punctuation, spelling, style, consistency, and other details.

corollary. A direct or easy consequence of a lemma, theorem, or proposition.

CTAN. The Comprehensive TEX Archive Network: a network of servers that hold up-to-date copies of public-domain versions of TEX, LATEX, and related packages and programs.

DOI. Digital Object Identifier. A unique string that permanently provides a link to an object.

Festschrift (or festschrift). [German] A collection of writings published in honor of a scholar.

GIF. Graphics Interchange Format. GIF was developed in 1987 by the CompuServe, an internet service provider, for diminishing image and animation size.

GNU. A recursive acronym for "Gnu's Not Unix," referring to a free UNIX compatible operating system.

hard copy. A printed copy of a file (document, book, etc.).

Harvard system. A system of citation by author name and year, for example, "see Knuth (1986)."

hybrid journal. A subscription journal that allows articles to be made open access on payment of an article processing charge (APC) by the authors.

hypothesis. A statement taken as a basis for further reasoning.

IMA. The Institute of Mathematics and Its Applications (Southend-on-Sea, UK). Also the Institute for Mathematics and Its Applications (based at the University of Minnesota, Minneapolis, USA).

impact factor. For a given journal and a given year, the number of citations in that year to articles published in the journal in the two preceding years divided by the number of articles published in the journal in the same two years.

Internet. The worldwide network of interconnected computer networks. It provides electronic mail, file transfer, news, remote login, and other services.

ISBN. International Standard Book Number.

ISSN. International Standard Serial Number.

JPEG. A lossy compression scheme for RGB images. It was developed in 1993 by the Joint Photographic Experts Group.

LaTeX. A macro package for TEX that simplifies the production of papers, books, and letters, and emphasizes the logical structure of a document. It permits automatic cross-referencing and supports bibliographies and indexing.

lemma. An auxiliary result needed in the proof of a theorem or proposition. May also be an independent result that does not merit the title theorem.

Linux. A family of open source Unix-like operating systems based on the Linux kernel, often using packages from the GNU project.

LMS. The London Mathematical Society (London, UK).

Lossless Compression. Compression of file size whereby the picture quality remains the same. Lossless files can be decompressed in order to revert the picture to its original condition.

Lossy Compression. Algorithms used to reduce file size by discarding the less important information. The general idea is to reduce the file size enough so that human perception of an altered picture file can be slightly or not at all detected. Lossy compression algorithms allow the user to determine resourceful ways to remove detail without humans noticing excessive modification.

MAA. The Mathematical Association of America (Washington, DC, USA).

macro. In computing, a shorthand notation for specifying a sequence of operations. For example, in typesetting this book in LATEX I used the definition \def\mw{mathematical writing} and typed \mw whenever I wanted the phrase "mathematical writing" to appear.

MakeIndex. A program that is used in the process of making an index for a LATEX document.

manuscript. Literally, a handwritten document. More generally, any unpublished document, particularly one submitted for publication.

Mathematical Reviews. A review publication run by the AMS and first published in 1940. Each listed paper is accompanied by a review or a reprint of the paper's abstract.

Mathematics Subject Classification. A classification scheme that divides mathematics into 64 sections numbered between 0 and 97, which are further divided into many subsections. An example of an entry is 65F05 (direct methods for solving linear systems).

offprint. See reprint.

open access publishing. A form of publishing in which publications are made freely available.

page charges. Charges levied by a publisher to offset the cost of publishing an article. In mathematics journals payment is usually optional.

PDF. Portable Document Format. A file format for representing documents developed by Adobe Systems and now an ISO standard. Can be read using the Adobe Acrobat software.

peer review. Refereeing done by peers of the author (people working in the same area). Should perhaps be called "peer refereeing," but "peer review" is standard.

PNG. A high-quality lossless compressed image file whereby the graphic can be altered by another person or whereby the image contains layers of graphics that must be kept separate from each other.

poster. A presentation in the form of one or more pieces of printed paper (or cloth), usually comprising text and graphics, attached to a poster board.

PostScript. A page description language developed by Adobe Systems, Inc. It preceded PDF, which is based on PostScript.

proceedings. A collection of papers describing the work presented at a conference or workshop. May also be included in the title of a journal: for example, Proceedings of the American Mathematical Society.

proofreading. The process of checking proofs for errors (usually by comparing them with an original) and (traditionally) marking the errors with standard proofreading symbols.

proofs. Typeset material ready for checking and correction.

proposition. Same meaning as theorem (but possibly regarded as a lesser result).

referee. A person who reviews a manuscript and advises an editor on its suitability for publication.

references. The list of publications cited in the text, or those publications themselves.

reprint. A separate printing of an article that appeared in a book or journal. A limited number may be supplied free of charge to the author.

reviewer. A person who reviews book proposals, journal submissions, or published books.

RGB: red green blue

running head. An abbreviated title that appears in the page header of pages in a published paper.

self-archiving. The practice whereby an author makes available online, on the author's own site or an appropriate repository, the author's own version of a paper accepted for publication in a journal.

SIAM. The Society for Industrial and Applied Mathematics (Philadelphia, Pennsylvania, USA).

technical report. A document published by an organization for external circulation, usually as part of a series.

TeX. A system for computer typesetting of mathematics, developed by Donald Knuth at Stanford University. Also used as a verb: "to TEX a paper" is to typeset the paper in TEX.

theorem. A major result of independent interest.

thesaurus. A list of words in which each word is followed by a list of words of similar meaning or sense. The main list may be arranged by meaning (Roget's Thesaurus) or alphabetically (most other thesauruses).

title. "The fewest possible words that adequately describe the contents of the paper" [113, Chap. 7].

Unix. A computer operating system developed at Bell Laboratories. Widely used on workstations and supercomputers.

URL. Uniform resource locator. The address of a resource on the internet.

widow line, widow word. A short last line of a paragraph appearing at the top of a page, or a word at the end of a paragraph appearing on a line by itself.

World Wide Web. One facet of the internet consisting of client and server computers handling multimedia documents. Term becoming obsolete.

Appendix A
COMMONLY CONFUSED WORDS

These are some of the pairs of words that are most often confused with each other. These words are often homonyms or homophones, but sometimes they are words that simply look similar or related but have entirely different meanings. Some of them are anagrams—words that have the same letters but are spelled differently (e.g., *depart* and *parted*)

adoptive / adopted: Children can be *adopted*, but parents are *adoptive*. In addition, a textbook for a curriculum can be *adopted*, not *adoptive*.

adapt / adopt: We *adapt* to something by getting used to it, but we *adopt* something by taking it in.

adverse / averse: An *adverse* situation is unfavorable, unpleasant, or difficult. To be *averse* means to be strongly against or opposed to something. It is often used in the negative, as in, "I am not *averse* to your idea . . ." which essentially means that the individual favors the idea.

affect / effect: affect means make a difference to, whereas effect means a result or bring about (a result).

affluent / effluent: Someone who is *affluent* is wealthy. The word *effluent* is often used as a noun and refers to sewage or liquid waste that is often discharged into a natural water environment such as a lake or ocean. Used as an adjective, *effluent* means flowing out.

allude / elude: To *allude* is to refer to something. To *elude*, on the other hand, is to avoid or escape something.

allusion / illusion: An *allusion* is a reference to something. An *illusion* is a perception that is not real or something seen that is interpreted by someone in one way, but in reality, it is something entirely different.

ambiguous / ambivalent: Something that is *ambiguous* has more than one meaning, is unclear in meaning, or is open to different interpretations, while someone who is *ambivalent* has mixed feelings about something.

amoral / immoral: *Amoral* means not concerned with morality or non-moral, while *immoral* describes actions with bad intentions, or not conforming to accepted standards of morality.

appraise / apprise: To *appraise* is to assess something, while to *apprise* means to inform.

ascent / assent: the word *ascent* is a noun that means the act of going up. To *assent* (v.) is to allow or agree to something.

atheist / agnostic: An *atheist* does not believe in one or more gods or divinities. An *agnostic* cannot prove or disprove the existence of one or more gods.

augur / auger: *Augur* is a verb which means be a sign of (as in a likely outcome), while the noun *auger* is a tool used for boring holes.

canvas / canvass: *Canvas* is a noun which refers to a closely knit fabric used for paintings and other art forms, sail masts for boats, or material used for tents. *Canvass* is a verb which means to solicit. People who *canvass* are usually those who ask for votes (politics) or those who try to sell merchandise (business).

categorize / classify: To categorize is to sort or group a type objects, events, or ideas by characteristics of conscious experience through conceptual discrimination. To classify is to categorize objects, events, or ideas hierarchically.

censure / censor: To *censure* is to express strong disapproval of something. To *censor*, on the other hand, means to suppress unacceptable parts of something, such as a book, film, public statement, and so forth.

cite / sight / site: To *cite* (v.) is to quote someone in writing or refer to an authority on a subject. *Sight* (n.) is the ability to see. A *site* is a place or location.

climactic / climatic: Something that is *climactic*, such as a good novel or play reaches a *climax* or high point. *Climatic* refers to climate as it relates to temperature and precipitation.

compare / contrast: People use these words interchangeably, but they have two distinct meanings. *Compare* means to examine something by identifying both similarities and differences. *Contrast*, as a verb, means to identify differences only.

complacent / complaisant: *Complacent* people are smug, self-serving, and often arrogant. Someone who is *complaisant*, on the other hand, is willing to please.

compare to / compare with: Use *compare to* when identifying either the similarities or the differences between two things. Use *compare with* when identifying both the similarities and the differences.

complement / compliment: Something *complements* something else by enhancing and contributing extra features to a given situation. A *compliment*, however, is an expression of praise or congratulatory remark.

continuous / continual: *Continuous* primarily means without interruption, and can refer to space as well as time, as in the cliffs form a continuous line along the coast; *continual*, on the other hand, typically means 'happening frequently, with intervals between', as in the bus service has been disrupted by continual breakdowns.

council / counsel: Council is an administrative or advisory body, while counsel is advice or guidance.

councilor / counselor: A *councilor* is a member of a council, whereas a *counselor* is someone who gives guidance on personal or psychological problems.

credible / creditable: Credible means believable, convincing, whereas creditable means 'deserving acknowledgement and praise'.

definite / definitive: Something that is *definite* is certain or possibly absolute. *Definitive* means decisive and with authority.

defuse / diffuse: To *defuse* is to remove a fuse from (such as an explosive device). It can also refer to reducing or lessening the danger or tension of something, such as a difficult situation. Something that is *diffuse* is dispersed or spread over a large region or area.

APPENDIX A

desert / dessert: A *desert* is a dry area or region. As a verb, *desert* means to leave someone or something. A *dessert* is the sweet course, usually at the end of a meal.

discreet / discrete: To be *discreet* is to avoid the possibility of attracting attention. Something that is *discrete* is separate or distinct from something else.

draw / drawer: Draw is used primarily as a verb, as in to sketch or to pull something out. Drawer means sliding storage compartment.

dual / duel: *Dual* is an adjective that refers to an object or situation with two components or attributes; A *duel* is a noun that refers to a combative fight.

egoism / egotism: It is egotism, not egoism, that means 'excessive conceit or self-absorption'; egoism is a less common and more technical word, for an ethical theory that treats self-interest as the foundation of morality.

elicit / illicit: To *elicit* means to draw out or get something (such as information). Something that is *illicit* is illegal or unlawful.

emigrate / immigrate: Someone *emigrates from* one country to another. Someone *immigrates to* a second country from a first.

envelop / envelope: *Envelop* without an e at the end means 'wrap up, cover, or surround completely', whereas an *envelope* with an e is a paper container used to enclose a letter or document.

exceptionable / exceptional: Exceptional means open to objection; causing disapproval or offence. Exceptional means not typical or unusually good.

farther / further: *Farther* refers to length or distance. *Further* means "to a greater degree," "additional," or "additionally." It refers to time or amount.

fawn / faun: A *fawn* is a young deer, and a light brown color. It can also be a verb which means *to cater to* or *to excessively accommodate*. A *faun* is a Roman deity that is part man, part goat.

flaunt / flout (v.): Flaunt means 'display ostentatiously', while flout means 'openly disregard (a rule)'.

flounder / founder (v.): Flounder generally means 'have trouble doing or understanding something, be confused', while founder means 'fail or come to nothing'.

APPENDIX A

forego / forgo (v.): Forego means precede, but is also a less common spelling for forgo, 'go without'.

great / grate: Something *great* is large, immense, excellent, or brilliant. A *grate* consists of cross bars over an opening (such as a sewer cover). To *grate* something is to irritate or shred it.

grisly / grizzly: As in *grizzly* bear *grisly* means 'causing horror or revulsion', whereas grizzly is from the same root as grizzled and refers to the bear's white-tipped fur.

hoard with horde: A hoard is a store of something valuable; horde is a disparaging term for a large group of people.

imply / infer: Imply is used with a speaker as its subject, as in "He implied that the General was a traitor," and indicates that the speaker is suggesting something though not making an explicit statement. "Infer" is used in sentences such as "We inferred from his words that the General was a traitor," and indicates that something in the speaker's words enabled the listeners to deduce that the man was a traitor.

incidence / incidents: An incidence is the rate of influence on something, such as "a high incidence of crime." Incidents are many occurrences or happenings.

incite / insight: To incite is to provoke or inflame something. Someone with insight has knowledge or a lot of experience.

instance / instants: Instance refers to a particular case or situation. An instant is a moment of time. So instants are two or more moments of time.

its / it's: The possessive its (as in turn the camera on its side) with the contraction it's (short for either it is or it has, as in it's my fault; it's been a hot day).

loath / loathe: Loath means reluctant or unwilling while loathe means 'dislike greatly'.

loose / lose: Loose means 'unfastened or set free', while lose means 'cease to have' or 'become unable to find'.

luxuriant / luxurious: Luxuriant means rich and profuse in growth, means 'characterized by luxury; very comfortable and extravagant'.

marital / martial: Marital refers to of marriage, while martial, 'of war'!

militate / mitigate: Militate, which is used in the form *militate against*, means to be an important factor in preventing something; *mitigate* means to make (something bad) less severe.

naturism and **naturist** with **naturalism** and **naturalist**: *Naturism* refers to the advocacy of a lifestyle that embraces physical exposure and a *naturist* is someone who follows naturism. *naturalism* is an artistic or literary approach or style; a *naturalist* is an expert in natural history, or an exponent of naturalism.

miner / minor: A *miner* works in mines (e.g., gold mines, coal mines, silver mines, etc.). A *minor* is too young to engage in activities that may be suitable for adults. As an adjective, *minor* means trivial or insignificant.

noisome / noisy (adj.): *Noisome* does not mean *noisy*. Rather, *noisome* means harmful or offensive (as in a very bad odor or a noxious smell). *Noisy* is the more common word, which describes something that is loud, boisterous, or raucous, or earsplitting.

notable / noticeable (adj.): *Notable* means worthy of distinction. *Noticeable* means attracting attention.

object / subject: An *object* is something to which something else, like a feeling or action, is directed. For example, an expensive piece of furniture might be the object of one's affection. A *subject* is a noun, or group of words that includes a noun, that acts on something (or someone) else in order to carry out an action.

officious / official (adj.): Someone who is *officious* asserts authority or interferes in an annoyingly domineering way. *Official* which means relating to an authority or public body and having the approval or authorization of such a body.

ordinance / ordnance (n.): An authoritative order, decree, or regulation is an *ordinance*. In contrast, *ordnance* is weaponry or munitions.

palate and palette (n.): The *palate* is the roof of the mouth; a *palette*, on the other hand, is an artist's board for mixing colors.

pedal / peddle / petal: *Pedal* is a noun denoting a foot-operated lever; as a verb it means to move by means of pedals. *Peddle* is a verb meaning to sell. The associated noun from pedal is pedaller (US pedaler), and the noun from peddle is peddler. A *petal*, however, is the often colorful part of a flower.

perquisite / prerequisite (n.): A *perquisite* is a special right or privilege enjoyed as a result of one's position; *prerequisite* is something that is required as a prior condition for something else; prerequisite can also be an adjective, meaning 'required as a prior condition'.

perspicuous / perspicacious (adj.): Something that is perspicuous is expressed clearly. Someone who is perspicacious has a ready understanding of things.

precede / proceed (v.): Precede means "come before." Proceed means to go forward (with something).

precept / percept (n.): A *precept* is a rule, teaching, or guideline. A *percept* is a perceived and often confusing or vague image.

principal / principle: As a noun a *principal* is the head of a school or organization. As an adjective, it is the main or chief thing. *Principle* is chiefly a noun and can mean a belief, law, or item of a code.

proscribe / prescribe (v.): We proscribe something by condemning it or forbidding it from occurring. Something is prescribed when it is recommended, usually by an authority. Physicians prescribe by issuing medicine to a patient.

regretful / regrettable (adj.): 'Feeling or showing regret' is regretful, with regrettable, which means 'giving rise to regret; undesirable'.

shear / sheer: 'Cut the wool off (a sheep)' is shear, with sheer, which as a verb means 'swerve or change course quickly' or 'avoid an unpleasant topic', and as an adjective means 'nothing but; absolute', 'perpendicular', or '(of a fabric) very thin'.

spatial / spacious (adj.): The word *spatial* means something having to do with space, area, or place. Something *spacious* is roomy.

stationary / stationery: *Stationary* is an adjective describing something that is unchanging or that stays in one place. *Stationery* is a noun denoting paper, writing materials, and other related office supplies.

story / storey (n.): A *story* is a tale or an account, while a *storey* is a floor of a building. In North America, the spelling *story* is sometimes used for *storey*.

titillate / titivate (adj.): *Titillate* means to excite. *Titivate* means to adorn or embellish, or make something look better.

tortuous / torturous (adj.): Something that is *tortuous* is full of twists and turns or excessively lengthy and complex. Something that is *torturous* is characterized by pain or suffering.

turbid / turgid (adj.): *Turbid* is generally used in reference to a liquid and means 'cloudy or opaque'; *turgid* tends to mean 'tediously pompous' or, in reference to a river, 'swollen, overflowing'.

unexceptionable / unexceptional (adj.): *Unexceptionable* describes something that cannot be taken exception to or inoffensive. *Unexceptional*, on the other hand, means not exceptional or ordinary.

unsociable / unsocial / antisocial (adj.): *Unsociable* means 'not enjoying the company of or engaging in activities with others'; *unsocial* usually means 'socially inconvenient' and typically refers to the hours of work of a job; *antisocial* means 'contrary to accepted social customs and therefore annoying'.

venal / venial (adj.): Someone who is *venal* is susceptible to bribery or corruptible. *Venial* is used in Christian theology in reference to sin (a venial sin, unlike a mortal sin, is not regarded as depriving the soul of divine grace).

vociferous / voracious (adj.): Someone who is *vociferous* is enthusiastic and determined in a loud way. Someone who is *voracious*, however, is avid or possibly gluttonous.

who's / whose: *Who's* is a contraction of *who is* or *who has*, while *whose* is a possessive pronoun, as in the following sentence: "Jerry is the one whose car broke down." In this case, *whose* refers to Jerry as the owner of the car that broke down.

wreath / wreathe: *Wreath* with no *e* at the end is a noun that means arrangement of flowers or green plants, while *wreathe* with an *e* is a verb meaning 'envelop, surround, or encircle'.

your / you're: *You're* is a contraction of *you are*, while *your* is a possessive determiner used in phrases such as *your turn*.

Using the Right Word

accept, except: The verb *accept* means "to receive"; the preposition *except* means "other than."

Susan graciously **accepted** defeat [verb]. All the boys **except** John were here [preposition].

are, hour, our: The plural form of the verb "to be" is *are*; an *hour* is a unit of time; the word *our* is a plural possessive pronoun.

Our friends *are* attending the meeting for an *hour*.

complement, compliment: *Complement* means "to complete or go with." *Compliment* is an expression of admiration or praise.

I **complimented** Aunt Anna by saying that her purse **complemented** her coat and dress.

fewer, less: *Fewer* refers to the number of separate units; *less* refers to bulk quantity.

There is *less* sand to play with, so we have *fewer* sandboxes to make.

good, well: *Good* is an adjective; *well* is nearly always an adverb.

The strange flying machines flew *well*. (The adverb *well* modifies *flew*.) They looked *good* as they flew overhead. (The adjective *good* modifies *they*.)

When used in writing about health, *well* is an adjective.

The runners did not feel *well* after the long, hard race.

loose, lose, loss: *Loose* (loos) means "free or untied"; *lose* (looz) means to "misplace" or "fail to win"; *loss* means "something lost."

Even though he didn't want to *lose* the *loose* tooth, it was no big *loss*.

past, passed: *Passed* is always a verb. *Past* can be used as a noun, as an adjective, or as a preposition.

That Mercedes-Benz *passed* my BMW [verb]. That old man won't forget the *past* [noun]. I'm sorry, but I'd rather not talk about my *past* life [adjective]. Gertrude walked *past* the cat and never saw it [preposition].

sit, set: *Sit* means "to put the body in a seated position." *Set* means "to place."

"How can you just *sit* there and watch as I *set* all these plates on the table?"

than, then: *Than* is used in a comparison; *then* tells when, for example, something happening after something else.

Then he cried and said that his big brother was bigger *than* my big brother. *Then* I cried.

their, there, they're: *Their* is a possessive pronoun, one which shows ownership. *There* is a pronoun used to point out a location. *They're* is the contraction for they are.

They're upset because ***their*** son dumped garbage over ***there***.

Who, Whom, and Whose: One of the most common difficulties in speech is to differentiate between *who* and *whom*. Before we learn how to identify the difference between these two words, let's get *whose* under control.

Whose

We use "whose" as a possessive pronoun that identifies an object or person in relationship with the subject. Take the following example.

Jake is the one *whose* homework was incomplete.

Who

The words who and whom are confusing. Fortunately, there is a mnemonic device that can help determine when to use *who* or *whom*. In short, mentally exchange the *who* or *whom* in your sentence with *they* or *them*. If *they* sounds correct, use *who*. If *them* sounds better, go with *whom*. We can make the parallel switches when writing with the singular personal pronoun—namely, she for who and her for whom and he for who and him for whom.

More technically speaking, *who* is the subject form and is used as the subject of the sentence. Use *who* when it serves as the subject of an adjective clause.

Whom

Whom is the object form and is used as a direct object, an indirect object, or an object of a preposition.

Appendix B

COMMONLY MISSPELLED WORDS AND SPELLING ASSESSMENT

Commonly Misspelled Words

Correct Spelling	Misspelling 1	Misspelling 2
1. accommodate	accomodate	acommodate
2. achieve	acheive	achive
3. acknowledgment	acknowledgement	
4. across	accross	
5. aggressive	agressive	
6. apparently	apparantly	
7. appearance	appearence	
8. argument	arguement	
9. basically	basicly	
10. beginning	begining	
11. believe	beleive	belive
12. bizarre[1]	bizzare	
13. business	buisness	bizness
14. calendar	calender	
15. Caribbean	Carribean	

Correct Spelling	Misspelling 1	Misspelling 2
16. cemetery	cemetary	
17. chauffeur	chauffer	
18. colleague	collegue	
19. coming	comming	
20. committee	commitee	commitie
21. completely	completly	
22. conscious	concious	contious
23. curiosity	curiousity	
24. definitely	definately	
25. desperate	desparate	
26. dilemma	dilemna	
27. disappear	dissapear	
28. disappoint	dissapoint	
29. ecstasy	ecstacy	
30. embarrass	embarass	
31. environment	enviroment	envirnment
32. existence	existance	
33. Fahrenheit	Farenheit	
34. familiar	familar	
35. finally	finaly	
36. fluorescent	florescent	
37. foreign	foriegn	
38. foreseeable	forseeable	
39. forty	fourty	
40. forward[2]	forward	
41. friend	freind	frend
42. gist	jist	
43. glamorous	glamourous	
44. government	goverment	
45. guard	gaurd	gard
46. happened	happend	

Correct Spelling	Misspelling 1	Misspelling 2
47. harass	harrass	
48. honorary	honourary	
49. humorous	humourous	
50. idiosyncrasy	idiosyncracy	
51. imitate	immitate	
52. immediately	immediatly	
53. incidentally	incidently	
54. independent	independant	
55. interrupt	interupt	
56. irresistible	irresistable	
57. judgment	judgement	
58. knowledge	knowlege	
59. liaise	liase	
60. lollipop	lollypop	
61. millennium	millenium	
62. Neanderthal	Neandertal	
63. necessary	neccessary	neccessery
64. noticeable	noticable	
65. occasion	ocassion	ocasion
66. occurred	occured	
67. occurrence	occurance	occurence
68. pavilion	pavillion	
69. persistent	persistant	
70. pharaoh[3]	pharoah	
71. piece[4]	peice	
72. politician	politition	
73. possession	posession	
74. preferred	prefered	
75. propaganda	propoganda	
76. publicly	publically	
77. really	realy	

Correct Spelling	Misspelling 1	Misspelling 2
78. receive	recieve	
79. referred	refered	
80. religious	Religious	
81. resistance	resistence	
82. sense	sence	
83. separate	seperate	
84. siege	seige	
85. successful	succesful	sucessful
86. supersede	supercede	
87. surprise	suprise	
88. tattoo	tatoo	
89. tendency	tendancy	
90. therefore	therefor	
91. threshold	threshhold	
92. tomorrow	tommorow	tommorrow
93. tongue	tungue	tounge
94. truly	truely	
95. unforeseen	unforseen	
96. unfortunately	unfortunatly	
97. until	untill	
98. weird	wierd	
99. wherever	whereever	
100. which	wich	

1 Bizarre and bazaar are homonyms, so the same could be applied to bazaar, an outdoor market usually associated with Middle Eastern cultures.

2 Forward and foreword are homonyms, so the same could be applied to foreword, an introductory section of a book that is usually written by someone other than the book's author as a celebratory endorsement of the book.

3 Pharaoh and farrow are homonyms, so the same could be applied to farrow, a verb which means to give birth to.

4 Piece and peace are homonyms, so the same could be applied to peace.

Directions: In each series below, select either the choice that is misspelled or choice "e" if all the words are spelled correctly.

1. a) noticeable b) courageous c) shyness d) arguement e) no mistakes
2. a) appraisal b) gradually c) proceed d) chief e) no mistakes
3. a) relieved b) seize c) shield d) famous e) no mistakes
4. a) character b) preferred c) establish d) omited e) no mistakes
5. a) national b) enclose c) confadential d) neighbor e) no mistakes
6. a) accidentally b) social c) trespass d) capable e) no mistakes
7. a) curious b) fiendish c) convertible d) dictater e) no mistakes
8. a) policy b) definately c) possess d) editor e) no mistakes
9. a) tomorrow b) worrying c) weird d) hemisphere e) no mistakes
10. a) beginning b) laboratory c) sheriff d) temperture e) no mistakes
11. a) acceptable b) diffrence c) bachelor d) refer e) no mistakes
12. a) ordinary b) embarrassed c) potatoe d) chauffeur e) no mistakes
13. a) lazyness b) secretary c) childhood d) dependent e) no mistakes
14. a) author b) colledge c) loyalty d) scholarship e) no mistakes
15. a) exercise b) recieve c) Wednesday d) safety e) no mistakes
16. a) prejudice b) vaxine c) generous d) vehicle e) no mistakes
17. a) volunteer b) video c) statue d) yield e) no mistakes

18. a) immitate   b) determine   c) decisions   d) audience   e) no mistakes
19. a) cataloge   b) approximate   c) repetition   d) salary   e) no mistakes
20. a) cautious   b) develope   c) general   d) incident   e) no mistakes
21. a) similiar   b) medicine   c) business   d) abundant   e) no mistakes
22. a) govener   b) bookkeeper   c) brilliant   d) traffic   e) no mistakes
23. a) twelfth   b) sovenir   c) strength   d) nucleus   e) no mistakes
24. a) library   b) refrence   c) preference   d) sincerely   e) no mistakes
25. a) successful   b) language   c) antonym   d) physical   e) no mistakes
26. a) symptoms   b) imaginary   c) admirable   d) invitation e) no mistakes
27. a) profesor   b) treaty   c) announce   d) pianos   e) no mistakes
28. a) neccesary   b) camouflage   c) truley   d) employer   e) no mistakes
29. a) miscellaneous b) miniature   c) athletic   d) humorous   e) no mistakes
30. a) marriage   b) tragedy   c) sophmore   d) height   e) no mistakes

APPENDIX B

Directions: In each series below, select either the choice that is misspelled or choice "e" if all the words are spelled correctly.

1. a) neurotic b) grammar c) envyous d) continuity e) no mistakes
2. a) dabble b) bazarre c) resign d) quiver e) no mistakes
3. a) guitar b) inhumane c) artillary d) framework e) no mistakes
4. a) bachelor b) mythology c) whether d) occult e) no mistakes
5. a) antelope b) casual c) decipher d) syncronize e) no mistakes
6. a) conversion b) melancholy c) gracious d) accessive e) no mistakes
7. a) phenomenan b) nocturnal c) remedial d) damaging e) no mistakes
8. a) champagne b) syanide c) amateur d) belligerent e) no mistakes
9. a) gradient b) meticulous c) persevere d) accommodate e) no mistakes
10. a) odometer b) fossill c) exaggerate d) depth e) no mistakes
11. a) dandruf b) beacon c) protrusion d) radar e) no mistakes
12. a) amphibian b) confidential c) dependency d) exponent e) no mistakes
13. a) ballast b) feudal c) classify d) numerater e) no mistakes
14. a) synthetic b) dedicate c) stationery d) oligarchy e) no mistakes
15. a) whisperring b) eruption c) tutorial d) vacuous e) no mistakes
16. a) immigrant b) unnatural c) illegible d) admireable e) no mistakes
17. a) meanness b) neutrall c) misapply d) argument e) no mistakes

Spelling

18.	a) comfortable	b) noticeable	c) excentric	d) reign	e) no mistakes
19.	a) prearrange	b) mischief	c) immortel	d) recede	e) no mistakes
20.	a) disuade	b) misinform	c) lofty	d) believable	e) no mistakes
21.	a) unnecessary	b) oxen	c) shelves	d) dispensce	e) no mistakes
22.	a) theory	b) prescribe	c) weight	d) boundery	e) no mistakes
23.	a) repercussion	b) doubtful	c) haphazard	d) agonizing	e) no mistakes
24.	a) collossal	b) faking	c) bouquet	d) explanation	e) no mistakes
25.	a) discipline	b) unanymous	c) weird	d) despondent	e) no mistakes
26.	a) rhthym	b) gourmet	) lieutenant	d) humorous	e) no mistakes
27.	a) freight	b) delectable	c) hypocriscy	d) possess	e) no mistakes
28.	a) ignorant	b) suspiscion	c) sponsor	d) originally	e) no mistakes
29.	a) leisure	b) medievel	c) seize	d) occurrence	e) no mistakes
30.	a) separated	b) magical	c) pheasant	d) labratory	e) no mistakes

APPENDIX B

Directions: In each series below, select either the choice that is misspelled or choice "e" if all the words are spelled correctly. There are 30 items in Section A.

1. a) noticeable | b) courageous | c) shyness | d) arguement | e) no mistakes
2. a) appraisal | b) gradually | c) proceed | d) chief | e) no mistakes
3. a) releived | b) seize | c) shield | d) famous | e) no mistakes
4. a) character | b) preferred | c) establish | d) omited | e) no mistakes
5. a) national | b) enclose | c) confadential | d) neighbor | e) no mistakes
6. a) accidentally | b) social | c) trespass | d) capable | e) no mistakes
7. a) curious | b) fiendish | c) convertible | d) dictater | e) no mistakes
8. a) policy | b) definate | c) possess | d) editor | e) no mistakes
9. a) tomorrow | b) worrying | c) weird | d) hemisphere | e) no mistakes
10. a) beginning | b) laboratory | c) sheriff | d) temperture | e) no mistakes
11. a) acceptable | b) diffrence | c) bachelor | d) refer | e) no mistakes
12. a) ordinary | b) embarrassed | c) potatoe | d) chauffeur | e) no mistakes
13. a) lazyness | b) secretary | c) childhood | d) dependent | e) no mistakes
14. a) author | b) colledge | c) loyalty | d) scholarship | e) no mistakes
15. a) exercise | b) recieve | c) Wednesday | d) safety | e) no mistakes
16. a) prejudice | b) vacine | c) generous | d) vehicle | e) no mistakes

17. a) volunteer   b) video   c) statue   d) yield   e) no mistakes
18. a) immitate   b) determine   c) decisions   d) audience   e) no mistakes
19. a) cataloge   b) approximate   c) repetition   d) salary   e) no mistakes
20. a) cautious   b) develope   c) general   d) incident   e) no mistakes
21. a) similiar   b) medicine   c) business   d) abundant   e) no mistakes
22. a) govener   b) bookkeeper   c) brilliant   d) traffic   e) no mistakes
23. a) twelfth   b) sovenir   c) strength   d) nucleus   e) no mistakes
24. a) library   b) refrence   c) preference   d) sincerely   e) no mistakes
25. a) successful   b) language   c) antonym   d) physical   e) no mistakes
26. a) symptoms   b) imaginary   c) admirable   d) invitation e) no mistakes
27. a) profesor   b) treaty   c) announce   d) pianos   e) no mistakes
28. a) neccesary   b) camouflage   c) truley   d) employer   e) no mistakes
29. a) miscellaneous b) miniature   c) athletic   d) humorous   e) no mistakes
30. a) marriage   b) tragedy   c) sophmore   d) height   e) no mistakes

APPENDIX B

Directions: In each series below, select either the choice that is misspelled or choice "e" if all the words are spelled correctly. There are 30 items in Section A.

1. a) neurotic b) grammar c) envyous d) continuity e) no mistakes
2. a) dabble b) bazarre c) resign d) quiver e) no mistakes
3. a) guitar b) inhumane c) artillary d) framework e) no mistakes
4. a) bachelor b) mythology c) whether d) occult e) no mistakes
5. a) antelope b) casual c) decipher d) syncronize e) no mistakes
6. a) conversion b) melancholy c) gracious d) excessave e) no mistakes
7. a) phenomenan b) nocturnal c) remedial d) damaging e) no mistakes
8. a) champagne b) syanide c) amateur d) belligerent e) no mistakes
9. a) gradient b) meticulous c) persevere d) accommodate e) no mistakes
10. a) odometer b) fossill c) exaggerate d) depth e) no mistakes
11. a) dandruf b) beacon c) protrusion d) radar e) no mistakes
12. a) amphibian b) confidential c) dependency d) exponent e) no mistakes
13. a) ballast b) feudal c) classify d) numerater e) no mistakes
14. a) synthetic b) dedicate c) stationery d) oligarchy e) no mistakes
15. a) whisperring b) eruption c) tutorial d) vacuous e) no mistakes
16. a) immigrant b) unnatural c) illegible d) admireable e) no mistakes

APPENDIX B

17. a) meanness — b) neutrall — c) misapply — d) argument — e) no mistakes
18. a) comfortable — b) noticeable — c) excentric — d) reign — e) no mistakes
19. a) prearrange — b) mischief — c) immortel — d) recede — e) no mistakes
20. a) disuade — b) misinform — c) lofty — d) believable — e) no mistakes
21. a) unnecessary — b) oxen — c) shelves — d) dispensce — e) no mistakes
22. a) theory — b) prescribe — c) weight — d) boundery — e) no mistakes
23. a) repercussion — b) doubtful — c) haphazard — d) agonizing — e) no mistakes
24. a) collossal — b) faking — c) bouquet — d) explanation — e) no mistakes
25. a) discipline — b) unanymous — c) weird — d) despondent — e) no mistakes
26. a) rhthym — b) gourmet —) lieutenant — d) humorous — e) no mistakes
27. a) freight — b) delectable — c) hypocriscy — d) possess — e) no mistakes
28. a) ignorant — b) suspiscion — c) sponsor — d) originally — e) no mistakes
29. a) leisure — b) medievel — c) seize — d) occurrence — e) no mistakes
30. a) separated — b) magical — c) pheasant — d) labratory — e) no mistakes

APPENDIX B

Directions: In each series below, select either the choice that is misspelled or choice "e" if all the words are spelled correctly. There are 30 items in Section A.

1. a) necktie   b) glimpse   c) lieutenent   d) Saturday   e) no mistakes
2. a) algebra   b) pronunciation   c) warrant   d) relief   e) no mistakes
3. a) tissue   b) benefited   c) typewriter   d) memarize   e) no mistakes
4. a) descent   b) completly   c) withhold   d) giraffe   e) no mistakes
5. a) secretive   b) interupt   c) international   d) courageous   e) no mistakes
6. a) unanimous   b) procedure   c) develope   d) siege   e) no mistakes
7. a) gasoline   b) resistence   c) embroidery   d) pageant   e) no mistakes
8. a) sophmore   b) senior   c) junior   d) freshman   e) no mistakes
9. a) schedule   b) permitted   c) endeavor   d) arguement   e) no mistakes
10. a) instalment   b) bargain   c) welfare   d) warrant   e) no mistakes
11. a) emergency   b) archtect   c) grievance   d) enthusiasm   e) no mistakes
12. a) aquire   b) accumulate   c) freight   d) hoping   e) no mistakes
13. a) cough   b) fourty   c) unnatural   d) picnic   e) no mistakes
14. a) bycycle   b) noticeable   c) tomorrow   d) foreign   e) no mistakes
15. a) scissors   b) secretary   c) seperate   d) peculiar   e) no mistakes
16. a) dissappoint   b) dishonest   c) distrust   d) disprove   e) no mistakes

17. a) pollute   b) pigeon   c) oxygen   d) medacine   e) no mistakes
18. a) fountain   b) library   c) liesure   d) lying   e) no mistakes
19. a) immigrant   b) emigrant   c) transparent   d) significant   e) no mistakes
20. a) limatation   b) commotion   c) observation   d) construction e) no mistakes
21. a) atom   b) combine   c) gravity   d) electricty   e) no mistakes
22. a) gnarled   b) Febuary   c) knowledge   d) kangaroo   e) no mistakes
23. a) phantom   b) paragraph   c) geography   d) triumph   e) no mistakes
24. a) mission   b) tradition   c) mesure   d) pleasure   e) no mistakes
25. a) kilometer   b) milimeter   c) bushels   d) gallon   e) no mistakes
26. a) symphony   b) requirement   c) fascinate   d) analysis   e) no mistakes
27. a) diffrence   b) irresistible   c) practice   d) sincerity   e) no mistakes
28. a) curiosity   b) abundant   c) apparatus   d) athalete   e) no mistakes
29. a) extraordinary b) questionnaire   c) tournement   d) syllable   e) no mistakes
30. a) secede   b) supercede   c) proceed   d) exceed   e) no mistakes

Directions: In each series below, select either the choice that is misspelled or choice "e" if all the words are spelled correctly. There are 30 items in Section A.

1. a) noticeable b) courageous c) shyness d) arguement e) no mistakes
2. a) appraisal b) gradually c) proceed d) chief e) no mistakes
3. a) releived b) seize c) shield d) famous e) no mistakes
4. a) character b) preferred c) establish d) omited e) no mistakes
5. a) national b) enclose c) confadential d) neighbor e) no mistakes
6. a) accidentally b) social c) trespass d) capable e) no mistakes
7. a) curious b) fiendish c) convertible d) dictater e) no mistakes
8. a) policy b) definate c) possess d) editor e) no mistakes
9. a) tomorrow b) worrying c) weird d) hemisphere e) no mistakes
10. a) beginning b) laboratory c) sheriff d) temperture e) no mistakes
11. a) acceptable b) diffrence c) bachelor d) refer e) no mistakes
12. a) ordinary b) embarrassed c) potatoe d) chauffeur e) no mistakes
13. a) lazyness b) secretary c) childhood d) dependent e) no mistakes
14. a) author b) colledge c) loyalty d) scholarship e) no mistakes
15. a) exercise b) recieve c) Wednesday d) safety e) no mistakes
16. a) prejudice b) vacine c) generous d) vehicle e) no mistakes

APPENDIX B

17. a) volunteer   b) video   c) statue   d) yield   e) no mistakes
18. a) immitate   b) determine   c) decisions   d) audience   e) no mistakes
19. a) cataloge   b) approximate   c) repetition   d) salary   e) no mistakes
20. a) cautious   b) develope   c) general   d) incident   e) no mistakes
21. a) similiar   b) medicine   c) business   d) abundant   e) no mistakes
22. a) govener   b) bookkeeper   c) brilliant   d) traffic   e) no mistakes
23. a) twelfth   b) sovenir   c) strength   d) nucleus   e) no mistakes
24. a) library   b) refrence   c) preference   d) sincerely   e) no mistakes
25. a) successful   b) language   c) antonym   d) physical   e) no mistakes
26. a) symptoms   b) imaginary   c) admirable   d) invitation e) no mistakes
27. a) profesor   b) treaty   c) announce   d) pianos   e) no mistakes
28. a) neccesary   b) camouflage   c) truley   d) employer   e) no mistakes
29. a) miscellaneous b) miniature   c) athletic   d) humorous   e) no mistakes
30. a) marriage   b) tragedy   c) sophmore   d) height   e) no mistakes

APPENDIX B

Directions: In each series below, select either the choice that is misspelled or choice "e" if all the words are spelled correctly. There are 30 items in Section A.

1. a) neurotic   b) grammar   c) envyous   d) continuity   e) no mistakes
2. a) dabble   b) bazarre   c) resign   d) quiver   e) no mistakes
3. a) guitar   b) inhumane   c) artillary d) framework   e) no mistakes
4. a) bachelor   b) mythology c) whether   d) occult   e) no mistakes
5. a) antelope   b) casual   c) decipher   d) syncronize   e) no mistakes
6. a) conversion b) melancholy c) gracious   d) excessave   e) no mistakes
7. a) phenomenan b) nocturnal c) remedial   d) damaging   e) no mistakes
8. a) champagne   b) syanide   c) amateur   d) belligerent   e) no mistakes
9. a) gradient   b) meticulous c) persevere d) accommodate   e) no mistakes
10. a) odometer   b) fossill   c) exaggerate d) depth   e) no mistakes
11. a) dandruf   b) beacon   c) protrusion d) radar   e) no mistakes
12. a) amphibian b) confidential c) dependency d) exponent   e) no mistakes
13. a) ballast   b) feudal   c) classify   d) numerater   e) no mistakes
14. a) synthetic b) dedicate   c) stationery d) oligarchy   e) no mistakes
15. a) whisperring b) eruption c) tutorial   d) vacuous   e) no mistakes
16. a) immigrant b) unnatural c) illegible d) admireable   e) no mistakes

APPENDIX B

17. a) meanness   b) neutrall   c) misapply   d) argument   e) no mistakes
18. a) comfortable   b) noticeable   c) excentric   d) reign   e) no mistakes
19. a) prearrange   b) mischief   c) immortel   d) recede   e) no mistakes
20. a) disuade   b) misinform   c) lofty   d) believable   e) no mistakes
21. a) unnecessary   b) oxen   c) sh-elves   d) dispensce   e) no mistakes
22. a) theory   b) prescribe   c) weight   d) boundery   e) no mistakes
23. a) repercussion   b) doubtful   c) haphazard   d) agonizing   e) no mistakes
24. a) collossal   b) faking   c) bouquet   d) explanation   e) no mistakes
25. a) discipline   b) unanymous   c) weird   d) despondent   e) no mistakes
26. a) rhthym   b) gourmet   ) lieutenant   d) humorous   e) no mistakes
27. a) freight   b) delectable   c) hypocriscy   d) possess   e) no mistakes
28. a) ignorant   b) suspiscion   c) sponsor   d) originally   e) no mistakes
29. a) leisure   b) medievel   c) seize   d) occurrence   e) no mistakes
30. a) separated   b) magical   c) pheasant   d) labratory   e) no mistakes

APPENDIX B

Directions: In each series below, select either the choice that is misspelled or choice "e" if all the words are spelled correctly. There are 30 items in Section A.

1. a) necktie — b) glimpse — c) lieutenent — d) Saturday — e) no mistakes
2. a) algebra — b) pronunciation — c) warrant — d) relief — e) no mistakes
3. a) tissue — b) benefited — c) typewriter — d) memarize — e) no mistakes
4. a) descent — b) completly — c) withhold — d) giraffe — e) no mistakes
5. a) secretive — b) interupt — c) international — d) courageous — e) no mistakes
6. a) unanimous — b) procedure — c) develope — d) siege — e) no mistakes
7. a) gasoline — b) resistence — c) embroidery — d) pageant — e) no mistakes
8. a) sophmore — b) senior — c) junior — d) freshman — e) no mistakes
9. a) schedule — b) permitted — c) endeavor — d) arguement — e) no mistakes
10. a) installment — b) bargain — c) welfare — d) warrant — e) no mistakes
11. a) emergency — b) archtect — c) grievance — d) enthusiasm — e) no mistakes
12. a)) aquire — b) accumulate — c) freight — d) hoping — e) no mistakes
13. a) cough — b) fourty — c) unnatural — d) picnic — e) no mistakes
14. a) bycycle — b) noticeable — c) tomorrow — d) foreign — e) no mistakes
15. a) scissors — b) secretary — c) seperate — d) peculiar — e) no mistakes
16. a) dissappoint — b) dishonest — c) distrust — d) disprove — e) no mistakes
17. a) pollute — b) pigeon — c) oxygen — d) medacine — e) no mistakes

APPENDIX B

18. a) fountain · b) library · c) liesure · d) lying · e) no mistakes
19. a) immigrant · b) emigrant · c) transparent · d) significant · e) no mistakes
20. a) limatation · b) commotion · c) observation · d) construction · e) no mistakes
21. a) atom · b) combine · c) gravity · d) electricty · e) no mistakes
22. a) gnarled · b) Febuary · c) knowledge · d) kangaroo · e) no mistakes
23. a) phantum · b) paragraph · c) geography · d) triumph · e) no mistakes
24. a) mission · b) tradition · c) mesure · d) pleasure · e) no mistakes
25. a) kilometer · b) milimeter · c) bushels · d) gallon · e) no mistakes
26. a) symphony · b) requirement · c) fascinate · d) analysis · e) no mistakes
27. a) diffrence · b) irresistible · c) practice · d) sincerity · e) no mistakes
28. a) curiosity · b) abundant · c) apparatus · d) athalete · e) no mistakes
29. a) extraordinary · b) questionnaire · c) tournement · d) syllable · e) no mistakes
30. a) secede · b) supercede · c) proceed · d) exceed · e) no mistakes

Appendix C

SOME GRAMMAR RULES AND SUGGESTIONS

Obtain a Grammar Book and <u>Learn the Parts of Speech</u>.

You should become familiar, as soon as you can, with the parts of speech, if you have not done so already. In specific, you need to know the following parts of speech, and how they fit within the context of a sentence.

Noun: sandstorm; California; car; book; Belinda; physics
Verb: go; sit; attempt; glide; juxtapose
Adjective: comfortable; blue; particular
Adverb: politely; too; nonetheless
Pronoun: him; her; it
Preposition: above; behind; for; by; toward
Conjunction: and; but; or
Interjection: Wow; Ouch; my goodness

Understand Your Tenses

Past tense means that the verb used is in the distant past, not recent past. For example: "My grandmother **went** to the park yesterday." Given that this event happened "yesterday," we use the past tense of "to go," which is "went." If the writer does not supply a word describing a period of time in the distant past (e.g., yesterday, last year, in 1884), then the following should be written: "My grandmother **has gone** to the park." Notice the use of the auxiliary verb "has."

Present tense never uses auxiliary verbs. Simply identify the verb with the correct ending for the present tense. For example: "My grandmother **goes** to the park."

Future tense almost always involves the auxiliary verb "will" and the infinitive form of the action verb. For example: "My grandmother **will go** to the park after dinner."

Diagram Sentences

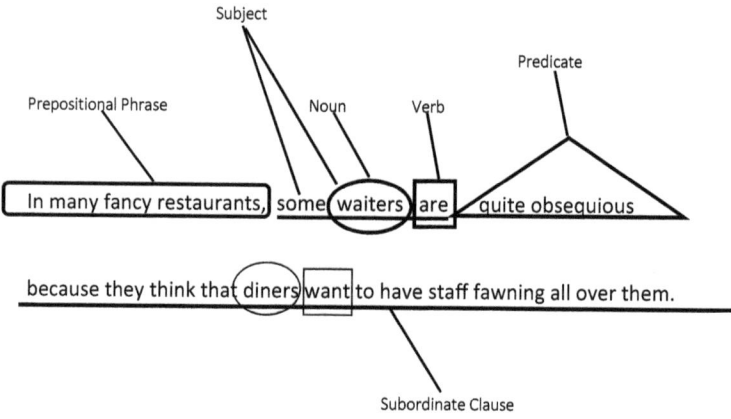

Verb Tenses

Use Present Tense

1. In statements of universal truth: I learned that water freezes at temperatures below zero degrees Celsius.
2. In statements about the contents of literature and other published works: In her article, Margaret Smith argues in favor of cognitively based instruction.

Use Past Tense

1. In statements concerning writing or publication of a book: Harriet Beecher Stowe wrote *Uncle Tom's Cabin* in 1851. The book was published a year later.

Use Present Perfect Tense

1. For an action that began in the past but continues to the present: Jack has been practicing law for the last 25 years.

Use Past Perfect Tense

1. For an earlier action that is mentioned in a later action: Carol ate the cake she had bought.
2. When using terms indicating something happening in the past (e.g., yesterday; in 1785; last month, etc.): Thomas Jefferson had signed the Declaration of Independence in 1776.

Use Future Perfect Tense

1. Or an action that will have been completed at a specific time in the future: By next year, I shall have completed my book.

Use a Present Participle

1. For an action that occurs at the same time as the verb: Running through the hallway, I heard the principal's voice.

Use the Perfect Participle

1. For an action that occurred before the main verb: Having listened to the guidelines, I then participated in the football game.

Use the Subjunctive Mood

1. To express a wish that is hypothetical or contrary to reality: If there were an outbreak, many people would be infected with the disease.

Rules of Subject-Verb Agreement

Subject-Verb Agreement refers to the grammatical correspondence between the subject and the verb of a sentence: I go; you go; she goes, etc.

1. **A verb must agree with its subject, not with any additional phrase in the sentence such as a prepositional or verbal phrase:** The *jar* with hundreds of marbles *is* on the mantle.
2. **In an inverted sentence beginning with a prepositional phrase, the verb still agrees with its subject:** In the cabinet *are* some *bags* of walnuts.
3. **A verb must agree with its subject, not its subject complement:** *Coffee and doughnuts are* her favorite snack. Her favorite *snack is* coffee and doughnuts.
4. **When a sentence begins with an existential subject-verb such as *there, here,* or *it*, the verb agrees with the subject, not the existential:** There *are* several *people* in the auditorium. There *is* a large construction *site* in the city.
5. **Indefinite pronouns such as *each, either, one, everyone, everybody,* and *everything* are singular:** *Everyone owns* a car in my neighborhood. *Each* student must *follow* the rules.
6. **Indefinite pronouns such as *several, few, both,* and *many* are plural:** *Few* dogs *live* a healthy life without humans.
7. **Indefinite pronouns such as *all, some, most,* and *none* may be singular or plural depending on their referents:** *Most* of the class *is* literate. *Some* of the students *are* struggling with verbal skills.
8. **Fractions such as *one-half* or *one-fifth* may be singular or plural depending on the referent:** *One-third* of the pie *is* finished. *One-third* of the pieces *are* remaining.
9. **Subjects joined by and take a plural verb unless the subjects are considered one item or unit:** *Frank* and *Annie are* in my social studies class. *Bacon and eggs is* Barney's favorite breakfast.

10. **In cases when the subjects are joined by or, nor, either ... or, or neither ... nor, the verb must agree with the subject closer to it:** Either the supervisor or her *subordinates are* responsible for the accident. Neither her subordinates nor the *supervisor is* responsible for the accident.
11. **Subjects preceded by *every, each,* and *many* are singular:** *Each* player *participates* in the playoffs. *Many a* mistake *has* to happen in order to improve.
12. **Relative pronouns, such as *who, which,* or *that,* which refer to plural antecedents require plural verbs. However, when the relative pronoun refers to a singular subject, the pronoun takes a singular verb:** Darren is one of the children who play on the swings. Darren is the only child who does not have an older brother.
13. **A collective noun, such as audience, faculty, jury, etc., requires a singular verb when the group is regarded as a whole, and a plural verb when the members of the group are regarded as individuals:** The *audience has* applauded after each performance. The *faculty are* meeting about their students.
14. **Subjects preceded by *the number of* or *the percentage of* are singular, while subjects preceded by *a number of* or *a percentage of* are plural:** *The number of* people in Charlie's focus group *remains* the same from one year to the next. *A number of* egrets *fly* to the Everglades each summer.
15. **Titles of books, companies, name brands, and groups are singular or plural depending on their meaning:** *The New York Yankees are* playing in Oakland. St. Augustine's *Confessions is* one of the most important literary works of late antiquity.
16. **Certain nouns of Latin and Greek origin have unusual singular and plural forms:** The *media are* dominating the spectacle. The *datum is* only one piece of information and therefore does not tell the reader very much.
17. **Some nouns such as *deer, shrimp,* and *sheep* have the same spellings for both their singular and plural forms. In these cases, the meaning of the sentences will determine whether they are singular or plural:** The *sheep are* grazing the pasture in large numbers. The strong wool merino *sheep is* with its offspring.
18. **Some nouns like *scissors, jeans,* and *wages* have plural forms but no singular counterparts. These nouns almost always take plural verbs:** The *scissors are* sharp.

19. **Some nouns ending in *–ics*, such as *economics*, *physics*, or *ethics*, take singular verbs:** *Phonics is* the most productive way to learn how to read. *Economics is* his favorite subject.
20. **Some nouns like *measles* and *news* appear to be plural but are actually singular:** The *news is* not new to Brenda.

Appendix D

LIST OF STEAM-RELATED STYLE GUIDES

Below, I have provided a selection of Editorial Style Guides for Academic Publication. Please note that while the list of style guides is comprehensive in the STEAM disciplines, it is by no means complete. That said, I have taken the liberty to include several of the more popular style guides that are used by STEAM authors. So, to this, I issue a mea culpa as I am solely responsible for any omission of a particular style guide that a reader might deem significant. I have also provided formats and examples of three types of publications in each of the style guides: journal article, book, and book chapter. I would suggest to readers to consult a particular style guide to 1) gain more experience using any particular one of the style guides, 2) identify formats of publication types other than journals, books, or book chapters, or 3) or learn formats of electronic publications. Finally, while the Modern Humanities Research Association (MHRA) Style Guide has the word "Humanities" in its title, this reference book is frequently used by engineers, architects, or by those in other domains in which technical writing is used. Regardless of any format, titles with proper nouns will always have an upper-case first letter. Any appearance of xx–xx indicate page numbers.

1. **American Chemical Society (ACS) Style Guide** (used by chemists and partially by physicists, biologists, and other disciplines in the natural sciences)

   Journal Article
   Last Name, First Initial.; Last Name, First Initial. Title. *Journal*, Year, *Volume* (issue), Pages.

   Example:
   Foster, J. C.; Varlas, S.; Couturaud, B.; Coe, J.; O'Reilly, R. K. Getting into Shape: Reflections on a New Generation of Cylindrical Nanostructures' Self-Assembly Using Polymer Building Block. *J. Am. Chem. Soc.* 2019, *141* (7), 2742–2753. DOI: 10.1021/jacs.8b08648

   Book
   Last Name, First Initial.; Last Name, First Initial. *Title*; Publisher, Year.

   Example:
   Frankel, F. *Picturing Science and Engineering*; MIT Press, 2018.

   Book Chapter
   Last Name, First Initial.; Last Name, First Initial. Title of Chapter. In *Title of Book*; Publisher, Year; pp. xx–xx.

   Example:
   Bard, A. J.; Faulkner, L. R. Double-Layer Structure and Absorption. In *Electrochemical Methods: Fundamentals and Applications*, 2nd ed.; John Wiley & Sons, 2001; pp 534–579.

2. **American Medical Association (AMA) Manual of Style: A Guide for Authors and Editors**
   Journal Article (used by physicians and researchers in the medical profession)
   Last Name First Initial, Last Name First Initial. Title of article: Subtitle of article. *Title of Journal* (abbreviated). Year: Volume(Issue):xx–xx.

   Example:
   Kleynhans J, Tempia S, Wolter N, et al; PHIRST-C Group. SARS-CoV-2 seroprevalence in a rural and urban household cohort during first and second waves of infections, South Africa, July 2020-March 2021. *Emerg Infect Dis.* 2021;27(12):3020–3029.

   Book
   Last Name First Initial, Last Name First Initial. *Title of Book: Subtitle of Book.* Publisher; Year.

   Example:
   Riegelman RK, Kirkwood B. *Public Health 101: Healthy People — Healthy Populations.* 2nd ed. Jones & Bartlett Learning; 2015.

   Book Chapter
   Last Name First Initial, Last Name First Initial. Title of book chapter: Subtitle of book chapter. In: Last Name First Initial, Last Name First Initial, eds. Title of Book: Subtitle of Book. Publisher; Year: xx–xx.

   Example:
   Bliss CM, Wolfe M. Chapter 34: Common clinical manifestations of gastrointestinal disease. In: Andreoli TE, Cecil RL, eds. *Andreoli and Carpenter's Cecil Essentials of Medicine.* 8th ed. Saunders/Elsevier; 2010:382–400.

3. **American Psychological Association (APA) Style Guide** (used by biologists, psychologists, education researchers, and other disciplines in the social sciences)

   Journal Article
   Last Name, First Initial., Last Name, First Initial., & Last Name, First Initial. (Year). Title of article. *Title of Journal, Volume*(Issue), pages.

   Example:
   Ginsburg, H. P., Lin, C. L., Ness, D., & Seo, K. H. (2003). Young American and Chinese children's everyday mathematical activity. *Mathematical Thinking and Learning, 5*(4), 235–258.

   Book
   Last Name, First Initial., Last Name, First Initial., & Last Name, First Initial. (Year). *Title of book: Subtitle of book.* Publisher.

   Example:
   Bransford, J. D., Brown, A. L., & Cocking, R. R. (2000). *How people learn: Brain, mind, experience, and school.* National Academy Press.

   Book Chapter
   Last Name, First Initial., Last Name, First Initial., & Last Name, First Initial. (Year). Title of chapter. In First Initial. Last Name & First Initial. Last Name (Eds.), Title of book: Subtitle of book (pp. xx–xx). Publisher.

   Example:
   Uttal, D. H., & Cohen, C. A. (2012). p? In B. Ross (Ed.), *Psychology of learning and motivation* (pp. 147–181). Academic Press.

4. **Chicago Manual of Style (used by researchers for general** and academic writing and publishing; if a specific style guide is used by a publisher, use the publisher's style guide of choice)

   Journal
   Last Name, First Name Middle Initial., First Name Middle Initial Last Name, and First Name Middle Initial Last Name. "Title of Article: Subtitle of Article." *Journal Title* Volume, no. Issue (Year).

   Example:
   Farenga, Stephen J., Daniel Ness, and Richard D. Sawyer. "Avoiding Equivalence by Leveling: Challenging the Consensus-Driven Curriculum that Defines Students as 'Average'." *Journal of Curriculum Theorizing* 30, no. 3 (2015).

   Book
   Last Name, First Name Middle Initial., First Name Middle Initial Last Name, and First Name Middle Initial Last Name. *Title of Book: Subtitle of Book*. Publisher, Year.

   Example:
   Ness, Daniel. *Block Parties: Identifying Emergent Steam Thinking Through Play*. Routledge, 2021.

   Book Chapter
   Last Name, First Name Middle Initial., First Name Middle Initial Last Name, and First Name Middle Initial Last Name. "Title of Book Chapter: Subtitle of Book Chapter." In *Title of Book: Subtitle of Book*, pp. xx–xx. Publisher, Year.

   Example:
   Blake, Brett Elizabeth. "A Broken Arch, a Broken Bridge, and a Broken Promise: Using Kincheloc's Critical Pedagogy Concepts to Teach about Race in an Urban Graduate School Classroom." In *Practicing Critical Pedagogy: The Influences of Joe L. Kincheloe*, pp. 121–130. Springer, 2016.

5. **Modern Humanities Research Association (MHRA) Style Guide**[1]
(used by engineers, architects, partially in mathematics, and disciplines in the humanities)

Journal

Author's first name and last name, 'Title of Article', *Title of Journal*, series number (if available), volume number.issue number (year of publication), page range, (page number referenced).

Example:

Luke Latario and others, 'Haves and Have-Nots: The State of Nanotechnology and STEM Education in US Baccalaureate Liberal Arts Colleges.' *Journal of Nano Education*, 6.1 (2014), 63–69.

Book

Author's first name and last name, *Title: Subtitle* (Place of Publication: Publisher, Year Published)

Example:

Sandra Schamroth Abrams, *Integrating Virtual and Traditional Learning in 6–12 Classrooms: A Layered Literacies Approach to Multimodal Meaning Making.* (New York: Routledge, 2014).

Book Chapter

Author's first name and last name, 'Title of chapter', in *Title: Subtitle*, ed. by Editors first name and last name (Place of Publication: Publisher, Year Published), page range of chapter, (page number cited).

Example:

Jana B. Milford, 'Environmental Law for Engineers', in Handbook of Environmental Engineering, ed. By M. Kutz (New York: Wiley, 2018), pp. 45–66.

6. **Council of Science Editors (CSE) Manual for Authors, Editors, and Publishers: Scientific Style and Format Style Guide** (used by biologists, chemists, physicists, geologists, and other researchers in the natural sciences)

   Journal
   Last Name First Initial, Last Name First Initial. Title of article: subtitle of article. Journal Title. Year: page numbers[2]

   Example:
   Farenga, SJ, Joyce, BA. Science sensibility. Sci Scope. 1999: 6–9.

   Book
   Last Name First Initial, Last Name First Initial. Title of book: subtitle of book. City(ST): Publisher; cYear.[3]

   Example:
   Ness D, Farenga SJ. Knowledge under construction: the importance of play in developing children's spatial and geometric thinking. Lanham (MD): Rowman & Littlefield; c2007.

   Book Chapter
   Last Name First Initial, Last Name First Initial. Title of book: subtitle of book. In: Last Name First Initial, editor. Title of book: subtitle of book. City(ST): Publisher; cYear. p. xx–xx.

   Example:
   Farenga SJ, Joyce BA, Ness D. Adaptive inquiry as the silver bullet: reconciling local curriculum, instruction, and assessment procedures with state-mandated testing in science. In: McMahon M, Simmons P, Somers R, DeBaets D, Crawley F, editors. Assessment in science: practical experiences and education research. Arlington(VA): National Science Teachers Association; c2006. p. 41–52.

Notes

1 This style guide is frequently used in the fields of architecture and engineering. See statement at the beginning of this Appendix.
2 Note that titles of journal articles in the natural sciences do not often contain subtitles.
3 (ST) refers to state or country, as in New York (NY) or London (UK). The "c" before the year stands for "copyright."

REFERENCES BY STEAM DISCIPLINE

General

Andrade, H. G. (2005). Teaching with rubrics: The good, the bad, and the ugly. *College Teaching, 53*(1), 27–31.

Arum, R., & Roksa, J. (2011). *Academically adrift: Limited learning on college campuses*. University of Chicago Press.

Ayers, W. (1993). *To teach: The journey of a teacher*. Teachers College Press.

Baszile, D. (2017b). On the virtues of currere. *The Currere Exchange Journal, 1*(1), vi–ix.

Blake, B. (1995). Broken silences: Writing and construction of cultural texts by urban preadolescent girls. *Journal of Educational Thought, 29*, 165–180.

Boaler, J. (2016). *Mathematical mindsets: Unleashing students' potential through creative math, inspiring messages, and innovative teaching*. Jossey-Bass.

Bromley, K., Irwin-DeVitis, L., & Modlo, M. (1995). *Graphic organizers: Visual strategies for active learning*. Scholastic Professional Books.

Chase, E., Morabito, N. P., & Abrams, S. S. (2020). *Writing in education: The art of writing for educators*. Brill.

Delandshere, G., & Petrosky, A. R. (1998). Assessment of complex performances: Limitations of key measurement assumptions. *Educational Researcher, 27*(2), 14–24.

Delandshere, G., & Petrosky, A. R. (1999). Anything can be measured, even colors can be measured: That's not the point. *Educational Researcher, 28*(6), 28–31.

Delandshere, G., & Petrosky, A. R. (2002). In a contact zone: Incongruities in the assessment of complex performances of English teaching designed for the National Board of

Professional Teaching Standards. In C. Dudley-Marling & C. Edlesky (Eds.), *Progressive language practices* (pp. 293–325). National Council of Teachers of English.

Elkonin, D. B. (2005). The psychology of play. *Journal of Russian and East European Psychology*, 43(1), 11–21, DOI: 10.1080/10610405.2005.11059245

Farenga, S. J., & Ness, D. (2017). SCALE down, SCALE back!: Academic freedom under siege through standards proliferation by para-educational enterprises. In D. Ness & S. J. Farenga (Eds.), *Alternatives to privatizing public education and curriculum* (pp. 78–98). Routledge.

Farenga, S. J., Ness, D., & Sawyer, R. D. (2015). Avoiding equivalence by leveling: Challenging the consensus-driven curriculum that defines students as "average". *Journal of Curriculum Theorizing* 30(3), 8–27.

Farenga, S. J., Ness, D., Johnson, B., & Johnson, D. D. (2010). *The importance of average: Playing the game of school to increase success and achievement*. Rowman & Littlefield.

Fleer, M. (2020). Engineering PlayWorld—a model of practice to support children to collectively design, imagine and think using engineering concepts. *Research in Science Education*, 1–16.

Freire, P. (1970/1996). *Pedagogy of the oppressed*. Continuum.

Groos, K. (1901). *The play of man*. Appleton.

Grunwald, E. (n.d.). The writing process: Making expository writing less stressful, more efficient, and more enlightening. https://writingprocess.mit.edu/

Henricks, T. (2020). Play studies: A brief history. *American Journal of Play*, 12(2), 117–155.

Hillocks, G. (1997). *How state mandatory assessment simplifies writing instruction in Illinois and Texas*. Paper presented at the annual meeting of the American Educational Research Association, Chicago, IL.

Hirsh-Pasek, K., & Golinkoff, R. M. (2008). Why play = learning. In R. E. Tremplay, R. G. Barr, R. D. V. Peters, & M. Boivin (Eds.), *Encyclopedia on early childhood development* (pp. 1-7). Centre of Excellence for Early Childhood Development.

Ho, A. D. (2008). The problem with "proficiency": Limitations of statistics and policy under No Child Left Behind. *Educational Researcher*, 37(6), 351–360.

Huizinga, J. (1955, originally published in 1938). *Homo ludens: A study of the play element in culture*. Beacon Press.

Illich, I. (1972). *Deschooling society*. Boyars.

Janks, H. (2013). Critical literacy in teaching and research. *Education Inquiry*, 4(2), 225–242.

Kimiecik, J. (2022). The feel of currere. *The Currere Exchange Journal*, 6(1), 63–73.

Kincheloe, J. (1998). Pinar's currere and identity in hyperreality: Grounding the post-formal notion of intrapersonal intelligence. In W. Pinar (Ed.), *Curriculum towards new identities* (pp. 129–142). Garland Press.

Koretz, D. M. (2009). *Measuring up: What educational testing really tells us*. Harvard University Press.

Lindqvist, G. (1995). *The aesthetics of play: A didactic study of play and culture in preschools*. Gotab.

Luke, A. (2012). Critical literacy: Foundational notes. *Theory into Practice*, 51(1), 4–11.

Mabry, L. (1999). Writing the rubric: Lingering effects of traditional standardized testing on direct writing assessment. *Phi Delta Kappan*, 80(9), 673–679.

McLaren, P., & da Silva, T. (2001). Decentering pedagogy: Critical literacy, resistance and the politics of memory. In P. McLaren & P. Leonard (Eds.), *Paulo Freire: A critical encounter* (pp. 47–89). Routledge.

Moser, J. S., Schroder, H. S., Heeter, C., Moran, T. P., & Lee, Y. H. (2011). Mind your errors: Evidence for a neural mechanism linking growth mind-set to adaptive post-error adjustments. *Psychological Science, 22*(12), 1484–1489.

Moskal, B., & Leydens, J. (2000). Scoring rubric development: Validity and reliability. *Practical Assessment, Research, and Evaluation, 7*(10), retrieved from http://pareonline.net/getvn.asp?v=7&n=10.

Piaget, J. (1951). *Play, dreams, and imitation in childhood.* C. Gattegno & F. M. Hodgson (Trans.). William Heinemann.

Pinar, W. (1975). Curerre: Toward reconceptualization. In W. Pinar (Ed.), *Curriculum theorizing: The reconceptualists* (pp. 396–414). McCutchan.

Pinar, W. (2004). *What is curriculum theory?* (1st ed.). Lawrence Erlbaum Associates.

Pinar, W. (2019). Currere. In J. Wearing, M. Ingersoll, C. DeLuca, B. Bolden, H. Ogden, & T. M. Christou (Eds.), *Key concepts in curriculum studies: Perspectives on the fundamentals* (pp. 50–52). Routledge.

Pinar, W. F., Slattery, P., Reynolds, W. M., & Taubman, P. M. (1995). *Understanding curriculum: An introduction to the study of historical and contemporary curriculum discourses.* Peter Lang.

Sawyer, R. D. (2022). Re/membering curricular entanglements: A currere of the present-absent curriculum of a gay high school student. *Journal of Curriculum Theorizing, 37*(1), 23–38.

Sawyer, R. D., & Liggett, T. (2012). Shifting positionalities: A critical discussion of a duoethnographic inquiry of a personal curriculum of post/colonialism. *International Journal of Qualitative Methods, 11*(5), 628–651.

Smith, B. A. (2013). Currere and critical pedagogy: Thinking critically about self-reflective methods. *Transnational Curriculum Inquiry, 10*(2), 3–16.

Steinberg, S. R., & Kincheloe, J. L. (2018). About power and critical pedagogy. In J. L. Kincheloe & S. R. Steinberg (Eds.), *Classroom teaching: An introduction* (pp. 29–46). Peter Lang.

Sutton-Smith, B. (2009). *The ambiguity of play.* Cambridge, MA: Harvard University Press.

UNESCO. (2019). *From access to empowerment: UNESCO strategy for gender equality in and through education 2019–2025.* https://unesdoc.unesco.org/ark:/48223/pf0000369000.

Vygotsky, L. (1966[1933]). Play and its role in the psychological development of the child. Lecture, Leningrad Pedagogical Institute. *Problems of Psychology, 6,* 62–76.

Vygotsky, L. S. (1986[1934]). *Thought and language* (A. Kozulin, Trans.). MIT Press.

Wang, W. (2017). Currere, subjective reconstruction and autobiographical theory. *Transnational Curriculum Inquiry, 14*(1–2), 110–141. https://doi.org/10.14288/tci.v14i1-2.188678

Wang, W. (2020). *Chinese currere, subjective reconstruction, and attunement: When calls my heart.* Palgrave Macmillan.

Wang, W. (2022). Currere, psychic speech and teacher education. *The Currere Exchange Journal, 6*(2), 18–26.

Natural Sciences

Alley, M. (2018). *The craft of scientific writing* (4th ed.). Springer.
Baker, L. M. (2017). *Writing in the environmental sciences*. Cambridge University Press.
Brown, M. C., Conway, J., & Sorensen, T. D. (2006). Development and implementation of a scoring rubric for aseptic technique. *American Journal of Pharmaceutical Education, 70*(6). Article 133.
Coppens, K. (2016). *Creative writing in science: Activities that inspire*. NSTA Press.
Einstein, A. (1905). On the electrodynamics of moving bodies. *Annalen der Physik, 17*(10), 891–921.
Farenga, S. J. (2000). Literacy in Science, Technology, and the Language Arts. *The Science Teacher, 67*(2), 70–72.
Farenga, S. J., Ness, D., & Hutchinson, M. (2008). Developing an Awareness of Pet Stewardship. *Science Scope, 32*(2), 58–63.
Farenga, S. J., Joyce, B. A., & Ness, D. (2004). Creating junior ethologists. *Science Scope, 28*(1), 60–62.
Farenga, S. J., Joyce, B. A., & Ness, D. (2006). Adaptive inquiry as the silver bullet: Reconciling local curriculum, instruction, and assessment procedures with state-mandated testing in science. In M. McMahon & P. Simmons (Eds.), *Assessment in science: Practical experiences and education research* (pp. 41–52). NSTA Press.
Field, J. B., Graf, L., & Link, K. P. (1952). The effect of methylxanthines on the hypocoagulability induced by chloroform in the dog. *Blood, 7*(4), 445–453.
Krajcik, J. S., & Sutherland, L. M. (2010). Supporting students in developing literacy in science. *Science, 328*(5977), 456–459.
Madamwar, D., Garg, N., & Shah, V. (2000). Cyanobacterial hydrogen production. *World Journal of Microbiology and Biotechnology, 16*(8), 757–767.
Schultz, D. (2013). *Eloquent science: A practical guide to becoming a better writer, speaker, and atmospheric scientist*. Springer.
Shah, V., Garg, N., & Madamwar, D. (2000). Record of the cyanobacteria present in the Hamisar pond of Bhuj, India. *Acta Botanica Malacitana, 25*, 175–180.
Smith, S.D., Huxman, T.E., Zitzer, S.F., Charlet, T.N., Housman, D.C., Coleman, J.S., Fenstermaker, L.K., Seemann, J.R., & Nowak, R.S. (2000). Elevated CO_2 increases productivity and invasive species success in an arid ecosystem. *Nature, 408*, 79–82.

Technology

Cope, B., Kalantzis, M., & Abrams, S. S. (2017). Multiliteracies: Meaning-making and learning in the era of digital text. In F. Serafini & E. Gee (Eds.), *Remixing multiliteracies: Theory and practice from new London to new times* (pp. 35–49). Teachers College Press.

International Technology and Engineering Educators Association. (2020). *Standards for technological and engineering literacy: The role of technology and engineering in STEM education*. Author.

Mishra, S., & Iyer, S. (2015). An exploration of problem posing-based activities as an assessment tool and as an instructional strategy. *Research and Practice in Technology Enhanced Learning, 10*(1), 1–19.

Turner, K. H., Abrams, S. S., Katíc, E., & Donovan, M. J. (2014). Demystifying digitalk: The what and why of the language teens use in digital writing. *Journal of Literacy Research, 46*(2), 157–193.

Engineering

Bélanger, P. (2013). Landscape infrastructure: Urbanism beyond engineering. In S. N. Pollalis, D. Schodek, & A.Georgoulias (Eds.), *Infrastructure sustainability and design* (pp. 276–315). Routledge.

Kosky, P., Balmer, R., Keat, W., & Wise, G. (2015). Exploring engineering: An introduction to engineering and design, 4th ed. Academic Press.

McCall, M., Fillenwarth, G. M., & Berdanier, C. G. P. (2020). Quantification of disciplinary discourse: An approach to teaching engineering resumé writing. In L. E. Bartlett, S. L. Tarabochia, A. R. Olinger, & M. J. Marshall (Eds.), *Diverse approaches to teaching, learning, and writing across the curriculum: IWAC at 25* (pp. 113–134). University Press of Colorado.

Paretti, M. C., Eriksson, A., & Gustafsson, M. (2019). Faculty and student perceptions of the impacts of communication in the disciplines (CID) on students' development as engineers. *IEEE Transactions on Professional Communication, 62*(1), 27–42.

Van Emden, J., & Becker, L. (2018). *Writing for engineers* (Macmillan Study Skills). Red Globe Press.

Art/Architecture

Lange, A., & Lange, J. M. (2012). *Writing about architecture: Mastering the language of buildings and cities*. Princeton Architectural Press.

Ness, D., & Farenga, S. J. (2007). *Knowledge under construction: The importance of play in developing children's spatial and geometric thinking*. Rowman & Littlefield Publishers.

Ness, D. (2005). Mapping your way to geographic awareness: Part II. *Science Scope, 28*(4), 59–63.

Ness, D. (2022). *Block parties: Identifying emergent STEAM thinking through play*. Routledge.

Spector, T., & Damron, R. (2017). *How architects write*. New York: Routledge.

Mathematics

Anderson, S. E. (1990). Worldmath curriculum: Fighting Eurocentrism in mathematics. *The Journal of Negro Education*, 59(3), 348–359.

Bertsekas, D. (2002). *Ten simple rules for mathematical writing*. Massachusetts Institute of Technology, Cambridge, MA. [Online]. Available: https://www.mit.edu/~dimitrib/Ten_Rules.pdf

Fuehrer, S. (2009). *Writing in math class?: Written communication in the mathematics classroom*. Math in Middle Institute Partnership Action Research Project Report. University of Nebraska.

Ginsburg, H. (1989). *Children's arithmetic: How they learn it and how you teach it*. Pro-Ed.

Ginsburg, H. P., Lin, C. L., Ness, D., & Seo, K. H. (2003). Young American and Chinese children's everyday mathematical activity. *Mathematical Thinking and Learning*, 5(4), 235–258.

Halmos, P. R. (1970). How to write mathematics. *L'Enseignement Mathématique*, 16(2), 123–152.

Knijnik, G. (1993). An ethnomathematical approach in mathematics education: A matter of political power. *For the Learning of Mathematics*, 13(3): 23–26.

Knuth, D. E., Larrabee, T., & Roberts, P. M. (1989). *Mathematical writing*. Mathematical Association of America.

Krantz, S. G. (1997). *A primer of mathematical writing: Being a disquisition on having your ideas recorded, typeset, published, read, and appreciated*. American Mathematical Society.

Miller, J. E. (2015). *The Chicago guide to writing about numbers*. University of Chicago Press.

Thomson, W. (2001). *A guide for the young economist*. MIT press.

Vivaldi, F. (2014). *Introduction to mathematical writing*. The University of London.

INDEX

A

Abrams, S. S., 71, 117
abstract
    components and length, scientific writing, 61
    natural and social science-related disciplines, 54–5
    symbolism, mathematical writing, 92
    technical reports, 83–4
academic writing, 87–9
Alley, M., 46
analysis of covariance (ANCOVA), 56
analysis of variance (ANOVA), 56
Anderson, S. E., 92
Andrade, H. G., 110
appendices, technical reports, 85
architecture
    strength and durability, 2
    Vitruvian Trinity of, 2, 3
arts, 1–3, 13. 105
Arum, R., 115–16

aseptic technique scoring rubric, 110, 111
Ayers, W., 107

B

Baker, L. M., 2
Balmer, R., 63
"banking model" of education, 65
Becker, L., 80
behavioral conditioning, 9
Belanger, P, 87
Berdanier, C. G. P, 80
Bertsekas, D, 92
Blake, B., 107
Bloom's Taxonomy of Educational Objectives, 12
Boaler, J., 32
brainstorming process, critical STEAM writing, 31–44
    concern about delivery or tone, 33
    drafting, 40–2

editing, 43
graphic organizer, 33–6
moodle, 36–8
outlining and planning stage, 38–40
preliminarily rough outline, 33
revising, 42
struggling situations, 32
Bromley, K., 33
Brown, M. C., 110
business case document, 81–2

C

Charlet, T.N., 59
Chase, E., 117
clarification questions, 66
Coleman, J.S., 59
Common Core State Standards (CCSS), 108
compression, 28
concern about delivery/tone, 33
conscientization, 19
constructive feedback, 118
content
  email writing, 77–8
  reading and reviewing, 51
  technology, 67–71
Conway, J., 110
Cope, B., 71
Coppens, K., 57
critical literacy, 9–10
  development of, 10
  in science, 13
  in STEAM, 9–14
critical STEAM writing
  appropriate tone for the audience, 21–9
  brainstorming process, 31–44
  chapter activity, 29–30
  direct vs. indirect introduction, 21–3
  and gender, 20–1
  grammar and syntax, 27
  usage errors, 25–6
  verb tense, 24–5

  and voice, 23–4
critics, 109
"the culture of silence," 109
Currere, 119

D

da Silva, T., 10
data analysis, 56
Delandshere, G., 110
direct learning, 9
disinformation, 19, 20
Donovan, M. J., 71
drafting, 40–2

E

editing, 43
education
  assessment/evaluation, 106
  gender equality, 106
  gender equity, 106
Einstein, A., 36, 49
Elkonin, D. B., 39
email writing
  advantages, 74
  content, 77–8
  convenience, 74
  grammar and mechanics, 76–7
  greeting the recipient, 78
  international communication, 74
  proofreading or checking, 78–9
  read on screens, 74
  technical report, 77
  texting language, 79–80
  time and cost saving, 74
  tone, 75
engineering, 3, 7, 13, 15, 16, 44, 63, 71, 73–89, 92, 107
  aeronautical, 55–6
  architecture and natural sciences, 4
  civil, 28

computer software, 64
equations, technical reports, 85–6
equity, in STEAM and writing, 106–7
Eriksson, A., 74
experiential learning, 9

F

Farenga, S. J., 45, 48, 65, 108, 110, 169
Fenstermaker, L.K., 59
Field, J. B., 47, 48
Fillenwarth, G. M., 80
Fleer, M., 39
Freirean perspective, 10
Freire, Paulo, 10, 65, 109
  concept of cultural conscientization, 19
Fuehrer, S., 96
full-class analysis of writing, 118

G

Garg, N., 50
gender equality in education, 106
gender equity in education, 106
Ginsburg, H. P., 95
Golinkoff, R. M., 39
grammar and mechanics, email
  writing, 76–7
graphical user interface (GUI)
  application, 12
graphic organizer, 33–6
graphics, technical reports, 86
graphing, 33–4, 36, 84, 86, 101
graphs, technical reports, 86
Groos, K., 39
Grunwald, E., 36
Gustafsson, M., 74

H

Halmos, P. R., 92

Heeter, C., 32
Henricks, T., 39
Hillocks, G., 110
Hirsh-Pasek, K., 39
Ho, A. D., 112
Housman, D.C., 59
Huizinga, J., 39
Hutchinson, M., 48
Huxman, T.E., 59

I

ideographs, 8
Illich, Ivan, 108
informal writing, 74
inspection reports, 86
instructional and curricular qualities, 11
instructional sequence, 9
integration, 3, 4, 13, 15–16
international communication, email
  writing, 74
International Technology and
  Engineering Educators Association
  (ITEA), 63, 64
introduction section, 84
Irwin-DeVitis, L., 33

J

Janks, H., 10
Johnson, B., 65
Johnson, D. D., 65
Joyce, B. A., 65

K

Kalantzis, M., 71
Katic, E., 71
Keat, W., 63
Kincheloe, J., 65
Knijnik, G., 10

Knuth, D. E., 92
Koretz, D. M., 110
Kosky, P., 63
Krajcik, J. S., 13
Krantz, S. G., 92

L

Lange, A., 105
Lange, J. M., 105
Larrabee, T., 92
Lee, Y. H., 32
letter writing, 80–1
Leydens, J., 110
Lin, C. L., 95
Lindqvist, G., 39
Link, K. P., 47, 48
Luke, A., 10

M

Mabry, L., 110
Madamwar, D., 50
malinformation, 19–20
mathematical writing
  abstract symbolism, 92
  captions, 94
  counterexamples, 94
  dependency graph and flow, specific arguments, 93, 101
  development of, 95
  examples, 94
  farmer jones problem
    eighth grader's analysis, 98, 102
    first graders' analysis, 100, 104
    high school junior's analysis, 99, 103
  figures and diagrams, 94
  hierarchical development, 93
  ideas using pictures, 94
  layout and structure, mathematical segment, 93, 100
  layout and structure of segment, 93
  math arguments or positions, 93
  mathematical hierarchy of form, 93, 102
  mathematical symbolism, 95, 96
  PowerPoint presentation, 93
  problem-solving, 96–100
  reading and reviewing, 91–2
  referencing, 94
  standards, 100
  writers of mathematics, 93
mathematics standards, 100
McCall, M., 80
McLaren, P., 10
method section, technical reports, 84
microtonal music performances, 13
Miller, J. E., 91
Mishra, S., & Iyer, S., 65
misinformation, 10, 19, 20
Modlo, M., 33
Montessori, Maria, 12
moodle, 36–8
Morabito, N. P., 117
Moran, T. P., 32
Moser, J. S., 32
Moskal, B., 110
motivational and cognitive currency, 12, 13
multiple analysis of covariance (MANCOVA), 56

N

national standards, adherence to, 107–16
Nation at Risk in 1983, 109
natural and social science-related disciplines
  abstract, 54–5
  discussion/conclusion, 57–8
  introduction, research paper, 55–6
  methods, 56–7
  pre-writing process, 52–4
  research papers in, 51–8
  results, research paper, 57
  title of research paper, 54

Northwest Regional Educational Laboratory (NWREL) Writing Rubric, 112–15
Nowak, R.S., 59

O

oral language, 8
outlining and planning stage, 38–40

P

Paretti, M. C., 74
peer analysis, 117
Petrosky, A. R., 110
physics concepts, 13
Piaget, J., 12, 39
Pinar, W. F., 119, 120
PISA, 109
political advantage or control, 10
Polya, George, 66
    four-step method for problem-solving, 66
post-Marxist concept, 19
pre-writing process, 52–4
proofreading/checking, email writing, 78–9

Q

questioning norms, critical steam writing, 105–21

R

ratiocination by writing genre, 116
reading and reviewing science content, 51
read on screens, email writing, 74
revising, 42
Reynolds, W. M., 119, 120
Roberts, P. M., 92
Roksa, J., 115–16

Rousseau, Jean-Jacques, 12
rubric assessment, 110–15

S

Sawyer, R. D, 49, 110, 169
Schroder, H. S., 32
Schultz, D., 52
science content
    reading, 51
    reviewing, 51
science-related experiences, 45
science writing, 48
    vs. scientific writing, 48–51
scientific literacy, 45
scientific writing, 46–7
    audience, 58
    clarity, 58
    components and length of abstract, 61
    logical writing, 58
    precision, 58
    redundant phrases, 60–1
    vs. science writing, 48–51
    succinctness, 58
seed knowledge, 65
Seemann, J.R., 59
semiotic function, 9
Seo, K. H., 95
Shah, V., 50
Slattery, P., 119, 120
small-group analysis, 118
Smith, B. A., 88
Smith, S.D., 59
social and political mobility, 10
socioeconomic and political contradictions, 10
Sorensen, T. D., 110
standards-based assessment movement, 109
Standards for Technological and Engineering Literacy Defining the Role of Technology and Engineering in STEM Education (STEL), 63
STEAM subjects

by-products and outcomes, 18
differences and similarities, 14–18
findings, 17–18
research and analysis, 16–17
special symbols and terminologies, 14–16
structure and form, 16
Steinberg, S. R., 65
struggling situations, 32
submicroscopic agents to physical illness, 12
subordinated cultural groups, 10
subitizing, 29, 30
Sutherland, L. M., 13
Sutton-Smith, B., 39
Systema Naturae, 47

T

Taubman, P. M., 119, 120
technical notes, 82–6. *see also* technical reports
technical reports
  abstract, 83–4
  appendices, 85
  discussion section, 84
  email writing, 77
  equations, 85–6
  graphics, 86
  graphs, 86
  introduction section, 84
  method section, 84
  references page, 84–5
  title page, 83
technologies
  content, 67–71
  developing ideas, 64–7
tension, 28

text and visual representations, 13
texting language, email writing, 79–80
Thomson, W., 15
TIMSS, 109
tone
  concern about delivery, 33
  email writing, 75
Turner, K. H., 71

U

United Nations Educational, Scientific and Cultural Organization (UNESCO), 106
unstructured interviews, 57
utilitas, 2

V

Van Emden, J., 80
venustas, 2, 3
Vitruvian Trinity of Architecture, 2, 3
Vivaldi, F., 97
Vygotsky, Lev, 12, 39

W

Wang, W., 119, 120
Wise, G., 63
writing by emulating models, 116–17
written communication, forms, 9

Z

Zitzer, S.F., 59

Critical Literacies and Language: Pedagogies of Social Justice

Lead Editor
Brett Elizabeth Blake

Editor
Judith M. Dunkerly

One of the most fundamental aspects of a just society is the right to create equitable and inclusive spaces of belonging for all people while also confronting injustice and oppression. However, we are now in a time where seeking justice and equity is met with neoliberalism, which pervades the academy at all levels of education. Yet, for many, this is not a time for retreat, but rather a moment of solidarity, a time to create new knowledge and understanding through struggle. As Freire wrote, "Knowledge emerges only through invention and re-invention, through the restless, impatient, continuing, hopeful inquiry human beings pursue in the world, with the world, and with each other." Thus, the purpose of this series is to provide literacy and language researchers, practitioners, as well as community activists, with a space to actualize and embody a restless, impatient never-finished objective of critical literacies and language education. It is the aim of this series to create a space to share research that promotes pedagogies of equity. We also recognize that different audiences have different needs. To that end, we seek to provide, when applicable, a "notebook" as a companion to research volumes to facilitate actionable steps for the PK-12 classroom or community spaces. This series is different as it approaches the dissemination of critical work from a place of intentionality to address the gap in disseminating research (typically read by scholars) and the need to have it "on the ground" for classroom teachers, community activists, and workers. By creating companion volumes (where applicable), there is a greater chance for sustained criticality in literacy education.

For additional information about this series or for the submission of manuscripts, please contact:

    Brett Elizabeth Blake, General Editor
    blakeb@stjohns.edu

To order other books in this series, please contact our Customer Service Department:

    peterlang@presswarehouse.com (within the U.S.)
    orders@peterlang.com (outside the U.S.)

Or browse online by series:

    www.peterlang.com

www.ingramcontent.com/pod-product-compliance
Ingram Content Group UK Ltd.
Pitfield, Milton Keynes, MK11 3LW, UK
UKHW021327180426
11947UKWH00017B/1477